Lecture Notes in Civil Engineering

Volume 122

Lecture Notes in Civil Engineering (LNCE) publishes the latest developments in Civil Engineering - quickly, informally and in top quality. Though original research reported in proceedings and post-proceedings represents the core of LNCE, edited volumes of exceptionally high quality and interest may also be considered for publication. Volumes published in LNCE embrace all aspects and subfields of, as well as new challenges in, Civil Engineering. Topics in the series include:

- Construction and Structural Mechanics
- Building Materials
- Concrete, Steel and Timber Structures
- Geotechnical Engineering
- Earthquake Engineering
- Coastal Engineering
- Ocean and Offshore Engineering; Ships and Floating Structures
- Hydraulics, Hydrology and Water Resources Engineering
- Environmental Engineering and Sustainability
- Structural Health and Monitoring
- Surveying and Geographical Information Systems
- Indoor Environments
- Transportation and Traffic
- Risk Analysis
- Safety and Security

To submit a proposal or request further information, please contact the appropriate Springer Editor:

- Pierpaolo Riva at pierpaolo.riva@springer.com (Europe and Americas);
- Swati Meherishi at swati.meherishi@springer.com (Asia - except China, and Australia, New Zealand);
- Wayne Hu at wayne.hu@springer.com (China).

All books in the series now indexed by Scopus and EI Compendex database!

More information about this series at http://www.springer.com/series/15087

Zbigniew Zembaty · Damian Beben ·
Zbigniew Perkowski · Adam Rak ·
Giovanni Bosco · Pranshoo Solanki
Editors

Environmental Challenges in Civil Engineering

Springer

Editors
Zbigniew Zembaty
Opole University of Technology
Opole, Poland

Zbigniew Perkowski
Opole University of Technology
Opole, Poland

Giovanni Bosco
University of L'Aquila
L'Aquila, Italy

Damian Beben
Opole University of Technology
Opole, Poland

Adam Rak
Opole University of Technology
Opole, Poland

Pranshoo Solanki
Construction Management Department
of Technology
Illinois State University
Normal, IL, USA

ISSN 2366-2557 ISSN 2366-2565 (electronic)
Lecture Notes in Civil Engineering
ISBN 978-3-030-63878-8 ISBN 978-3-030-63879-5 (eBook)
https://doi.org/10.1007/978-3-030-63879-5

This Springer imprint is published by the registered company Springer Nature Switzerland AG
The registered company address is: Gewerbestrasse 11, 6330 Cham, Switzerland

Scientific Committee of ECCE 2020 Conference

Organizing Committee of ECCE 2020 Conference

Chairman

Dr. Wiesław Baran, Opole University of Technology/PACET—Opole Branch, Poland

Members

Prof. Adam Rak, Opole University of Technology/Opole District Chamber of Civil Engineers, Poland
Prof. Zbigniew Perkowski, Opole University of Technology/Katowice Branch of Polish Academy of Sciences—Commission of Civil Engineering, Poland
Dr. Dominika Pilarska, Opole University of Technology, Poland
Dr. Jan Centkowski, PACET—Opole Branch, Poland
Mr. Jan Broniewicz, Opole District Chamber of Civil Engineers, Poland
Krzysztof Irek, M.Sc., Opole University of Technology, Poland
Piotr Nowacki, M.Sc., Opole University of Technology, Poland

Preface

The presented book *Environmental Challenges in Civil Engineering* with its individual 16 chapters contains the results of interdisciplinary scientific research in the field of civil engineering (especially construction and materials) together with elements of architecture as well as environmental protection engineering and sustainable development. An important element of each chapter is an analysis of selected literature issues together with a description of the authors' research tools as well as conducted experiments. The editors hope that these aspects of the work may contribute to creating more synergies between practice and research in order to increase the development and commercialization of research in the field of advanced construction technologies and materials. Other words, it can be recommended not only to scientists, but also to designers, contractors, practitioners involved in the design of building objects as well as employees of architectural and building administration and construction supervision. Moreover, it can be interesting for companies promoting modern technologies and products in the field of civil engineering, as well as activities related to reducing energy consumption in construction while maintaining the principles of sustainable development.

Among others, such important issues considered in the work should be distinguished:

- designing of buildings, their elements, and transport infrastructure, including durability and protection or environmental aspects of sustainable construction, reconstruction and refurbishment of historic buildings,
- some modern methods and innovative technologies in material engineering and building physics,
- architecture and urban planning in the context of the idea of sustainable development,
- BIM in construction,
- economic, and organizational issues of preparation and implementation of construction projects.

The chapters are scientific contributions sent by authors to the 4th International Conference *Environmental Challenges in Civil Engineering 2020* (ECCE 2020).

Finally, based on the independent reviews, the editors selected 16 works to the publication. The conference was organized by Polish Association of Civil Engineers and Technicians (PACET)—Opole Branch, Faculty of Civil Engineering and Architecture at Opole University of Technology, Opole District Chamber of Civil Engineers, and Commission of Civil Engineering at Katowice Branch of the Polish Academy of Sciences.

The prestige of the ECCE 2020 Conference is ensured by the 40-person Scientific Committee. The invitation to participate in the ECCE 2020 Scientific Committee was accepted by scientific authorities in the field of civil construction and environmental engineering not only in Poland, but also in nine European countries and the USA.

The conference was devoted first of all to the issues listed above. Unfortunately, due to the COVID-19 pandemic in 2020, direct participation of the authors during the Conference in Opole and also planned accompanying events with a technical dimension had to be cancelled. That is why the editors publishing this book and reaching a wider audience also have the personal satisfaction that this way they have a chance to increase the range of impact and share interesting research results sent by the authors despite the very difficult situation that has happened to all of us this year.

We are convinced that the published selected scientific works can extend international cooperation between national and foreign centres in the field of research, undertaking joint projects, creating cooperation networks and exchanging experiences.

It should be mentioned that ECCE 2020 had an international dimension and was a continuation of conferences previously organized in Opole, i.e. in 2014, 2016, and 2018, on the same issues. In particular, the papers from the 3rd ECCE Conference published by MATEC (indexed by Web of Science database) in 2018 can be also recommended to readers as an interesting trace of past events from this series. This year, the organizers are opening a new ECCE history page by publishing the best papers via Springer, for which they express their great gratitude.

The editors would also like to thank everyone who helped to organize the ECCE 2020 Conference during this difficult time and publish this book. Special thanks in this regard go to the members of scientific and organizing committees and the reviewers of chapters.

Opole, Poland — Zbigniew Zembaty
Opole, Poland — Damian Beben
Opole, Poland — Zbigniew Perkowski
Opole, Poland — Adam Rak
L'Aquila, Italy — Giovanni Bosco
Normal, USA — Pranshoo Solanki

Contents

About the Editors

Zbigniew Zembaty, Ph.D., D.Sc. Professor. E-mail: z.zembaty@po.edu.pl. He has been working since 1980 at the Faculty of Civil Engineering and Architecture of Opole University of Technology (Poland) where he is currently the Dean and Head of Department of Structural Engineering and Mechanics. He is an author or co-author of a book and about 100 other publications including 43 covered by Science Citation Index (core collection) with 327 citations and respective h-index 14. The main areas of scientific research cover civil engineering structural dynamics, random vibrations, inverse problems, seismic engineering, geophysics. Co-editor of two Springer books: *Seismic Behaviour and Design of Irregular and Complex Civil Structures II* (2016), *Seismic Behaviour and Design of Irregular and Complex Civil Structures III* (to appear in 2020). Cooperation with Milan Polytechnic, Trento University (Italy) and Israel Institute of Technology, including frequent participation in the respective doctoral schools with lectures on random vibrations. Participation in scientific committees of numerous international conferences including World Conference on Earthquake Engineering (16th WCEE in Santiago, Chile in 2017). Actual active research in investigating rotational component of seismic ground motion. Member of special int. group IWGoRS which deals with introducing the so-called 6-component seismology and seismic engineering.

Damian Bęben, Ph.D., D.Sc. Associate Professor. E-mail: d.beben@po.edu.pl. He is the Chairman of the scientific council of the civil engineering and transport discipline at the Faculty of Civil Engineering and Architecture, Opole University of Technology (Poland). An author and co-author of three books and over 180 publications on the national and international conferences as well as in the peer-reviewed scientific journals indexed in the Journal Citation Reports. He has an h-index of 12, according to the Web of Science Core Collection. He is a reviewer in many scientific international journals and the National Centre for Research and Development; Member of the International Association for Bridge Maintenance and Safety (IABMAS); International Association for Life-Cycle Civil Engineering (IALCCE); International Association of Computer Science and Information

Technology (IACSIT); Transportation Research Board (TRB) of the National Academies; Committee on Subsurface Soil-Structure Interaction (AFS40). He was the scholarship holders of the Foundation of Polish Science for the young prominent scientists; scientific scholarship for outstanding young scientists awarded by the Ministry of Science and Higher Education; 2011 Outstanding Reviewer for the Journal of Bridge Engineering (ASCE); and the European Social Found for Ph.D. Research interests: analysis of soil-steel bridge, durability of engineering structures, field load tests of structures, non-destructive evaluation of structures, environmental protection in transportation engineering.

Zbigniew Perkowski, Ph.D., D.Sc. Associate Professor. E-mail: z.perkowski@po.edu.pl. He has been working since 1999 at the Faculty of Civil Engineering and Architecture of Opole University of Technology (Poland) where he is currently the Head of the Mechanics of Materials Team in the Materials Physics Department and the Faculty Editor. He is an author or co-author of three books, and over 60 publications including proceedings of the domestic and international conferences as well as articles in the peer-reviewed scientific journals including: *International Journal of Heat and Mass Transfer, Journal of Building Physics, Engineering Structures, Materials*. He was a member of scientific boards of two international conferences and belongs to two sections of the Committee of Civil Engineering and Hydroengineering of the Polish Academy of Sciences—Section of Building Physics and Section of Mechanics of Structures and Materials. He is a Chairman of the Commission of Civil Engineering at Katowice Branch of Polish Academy of Sciences in the terms of office 2015–2018 and 2019–2022. In 2013, he received the Professor W. Żenczykowski's Award of Polish Association of Civil Engineers and Technicians (PACET) for the achievements in research in the field of civil engineering concerning damage mechanics. The most important research interests: damage mechanics, mechanics of concrete, composite structures, building physics, and ultrasonic testing of concrete structures.

Adam Rak, Ph.D., D.Sc., Associate Professor. E-mail: a.rak@po.opole.pl. He is the Head of Department of Civil Engineering and Construction Processes at the Faculty of Civil Engineering and Architecture, Opole University of Technology (Poland). Research activities focus on conducting researches on the application of modern mathematical methods in the modelling of technological processes in environmental engineering and construction, construction of numerical models to optimize the procedures of supervision and control of investment processes, analysis of the issues of engineering and environmental protection in construction, environmental conditions of preparation and the implementation of investment and the problems of processes that shape the quality of surface and underground waters. Author or co-author of more than 100 studies, concepts, designs and expertise in the field of water and sewage management, agricultural waste management and environmental impact assessment. The current scientific achievements include 185 publications and studies. Forty-four of these works have been published, including two books, editing of four multi-authorial monographs, 31 original works. Some of the works were published such scientific journals as: *Polish Journal of Environmental Studies, International Journal of Engineering Research and Application, International Journal of Water Sciences, Open Journal of Architectural Design, Water Supply—Water Practice and Technology, Wulfenia Journal*, and as chapters in monographs: *Cutting Edge Research in New Technologies and Environmental Engineering III*—Taylor & Francis Group.

Giovanni Bosco, Ph.D. Associate Professor. E-mail: giovanni.bosco@univaq.it. He works at the Department of Civil Engineering, Construction and Architectural Engineering, Environmental Engineering, Università degli Studi dell'Aquila, Italy. He is graduated in Civil Engineering "cum laude" at the University of Palermo in 1982. Master of Science in Civil Engineering, University of California at Berkeley, USA, in 1984 with a thesis on selection of soil properties for use in dynamic analyses and evaluation of liquefaction potential of natural sand deposits. He is an author and co-author of about 40 conference papers, papers in international journals, research reports, on various topics including: the use and interpretation of situ testing to evaluate stiffness and strength of soils and soil properties for use in dynamic response analyses, modelling the mechanical behaviour of unsaturated soils on the basis of laboratory testing and analyses of real slopes in unsaturated soils, investigation on the long-term response of sand via laboratory testing, modeling the response of earth reinforced soils based on laboratory investigations on the response of soils-structure interfaces by means of two-dimensional large-scale apparatus, modeling the response of piled foundations by means of finite element analyses of the homogenized medium (piles and soils), design of drainage systems to improve the stability of earth slopes, selection of soil properties for use in dynamic analyses, modeling toppling failure of rock slopes taking into account complex sliding and overturning kinematic failure modes. He is a member of various Eurocodes working groups and project teams related to the preparation of the CEN/TC 250/SC 7 Eurocode 7: Geotechnical Design, General Rules and the CEN/TC 250/SC 7 Eurocode 7, Part 3: Geotechnical Design, Design Assisted by Field Testing. As a professional engineer in civil engineering, he have been involved in more than 400 project in Italy and abroad, including geotechnical design and consulting at the design stage and during construction.

Pranshoo Solanki, Ph.D., D.Sc. Associate Professor. E-mail: psolank@ilstu.edu. He has been working at the Department of Technology at Illinois State University. He has over 10 years of academic and professional experience in the field of civil and geotechnical/pavement engineering. He received his doctorate in Civil Engineering from the University of Oklahoma and master's degree in Civil Engineering from Indian Institute of Technology Delhi. His research efforts have resulted in over 60 publications and over 30 research presentations in a number of national/international conferences. Dr. Solanki worked on a number of transportation and construction projects funded by agencies such as Illinois Transportation Center and Environmental Protection Agency. He has been recognized for his research work through awards at university and professional level.

Dr. Solanki is a registered Professional Engineer (PE) in the USA and serving as a member of Transportation Research Board (TRB), American Society of Civil Engineers (ASCE), American Concrete Institute (ACI) and American Society of Engineering Education (ASEE). Dr. Solanki continues to teach courses in the area of geotechnical engineering, construction materials and design at Illinois State University.

Innovative Bridge Concept

J. Bujnak, P. Bujnakova, and J. Odrobiňák

Abstract The design of high speed railway lines assumes overall geometrical parameters with proper ballasted track bed and innovative bridge structure arrangement. The specific provision for evaluation of dynamic effects, riding comfort requirement, thermal forces actions, and approach arrangement are subjects of this paper. Testing methods for verification of global behaviour of bridges are also presented. The computer modelling using more sophisticated analyses is discussed and experimental assessment given. Typical examples are described.

Keywords Railway corridor · Bridge types · Track interaction · Ballastless track

1 Bridge Particularities for Higher Operational Velocity

Europe is decidedly supporting and financing international high speed links between its more relevant countries. The Baltic-Adriatic multimodal railway corridor, as a part of European TEN-T infrastructure, runs from North to South, connecting core Baltic ports with ports of the Adriatic Sea, and includes 4 285 km of 1435 mm standard gauge railway infrastructure [1]. On the Slovak part, actual actions cover reconstruction of national railway section of the line Zwardoń—Žilina—Bratislava—Wien for higher speed. In view of the restrictions which are imposed at the approaches to large towns, a running speed in excess of 160 km/h seems to be the reference level for this new railway line.

Introduction of higher speed by developing existing conventional rail systems and its greater velocity is primarily the fact of considerable infrastructure improvements. The main advantage of this retrofitting of conventional rail system is lower cost and much less time required in introducing a high-speed trains. Nevertheless, higher speed and heavy freight trains have different demands on track standards, concerning

J. Bujnak (✉) · P. Bujnakova · J. Odrobiňák
Faculty of Civil Engineering, University of Zilina, Zilina, Slovakia
e-mail: bujnak@uniza.sk

Z. Zembaty et al. (eds.), *Environmental Challenges in Civil Engineering*, Lecture Notes in Civil Engineering 122,
https://doi.org/10.1007/978-3-030-63879-5_1

horizontal alignment, cant, gradients and vertical curves. Specific technology and codes requirements should be applied in order to be taken into account rail-deck interaction, dynamic impact of live loads, fatigue-resistance design, aerodynamic effect, derailment or collisions.

Essential prerequisite for high speed corridor is to have control on the degradation of track geometry, so as to keep various tolerances well within the specified limits. Axle-load, speed, sub-grade characteristics and improvement of subgrade in poor ground areas and bridges is recognized as one of the most significant factor. Therefore, track bed for high-speed lines is obviously much deeper than in the case of conventional railways. With speed increasing, the noise problems appear. Its nature changes with the speed increasing. Measures in the forms of screens, mounds of earth should be taken to protect against noise, as well as possible modifications of the route or the creation of artificial tunnels or covered sections and eventual modifications to the rolling stock. Normally level crossing is not suitable for higher speed train operation and hence road overpasses or road under bridges need to be constructed. In addition, complete lines shall be screened to avoid interference of local people and animals. Consequently, analyses of bridges for higher speed tracks present several particularities.

1.1 Specific Railway Assessments

1.1.1 Traffic Actions on Bridge Structures

Besides assessments related to structures and materials, there are some specific railway criteria to be checked. The characteristic vertical live load models *LM 71* are given in EN 1991-2 [2]. A high speed train convoy can dynamically excite the deformations of the bridge deck with the result of a resonant growth of the deflections. So, high speed load model *HSLM* should be along with taken into account according to EN 1991-2 [2]. The authors of the documents [3] recommend similar approaches to the standard procedure. Both loading shall be multiplied by a factor α on lines carrying rail traffic which is heavier or lighter than normal rail traffic. When multiplied by the factor α, the loads came to be classified vertical loads. The value 1.21 is normally recommended on lines for freight traffic and international lines. On lines carrying rail traffic which is lighter, the factor α shall be 1.1.

Together with vertical actions, some horizontal forces due to rail traffic must be taken into account. Traction and braking forces act at the top of the rails in the longitudinal direction of the track. When the track is continuous at one or both ends of the bridge, only a proportion of these acceleration and braking forces is transferred through the deck to the bearings, the remainder of the forces are transmitted through the track where it is resisted behind the abutments. The additional but transversal centrifugal forces are considered fully transmitted through the deck to the bearings. The nosing forces, also lateral ones, have generally only local effects. The centrifugal and nosing forces, traction and braking actions shall be multiplied by the same factor α. But, all calculations for fatigue are done with the *LM 71* and the factor $\alpha = 1.0$.

The stresses, deformations and associated bridge deck accelerations induced in a bridge are increased and decreased under the effects of moving traffic. Quasi-static analysis can be usually carried out with classified vertical loads representing normal rail traffic. Resulting static load effects are then multiplied by the dynamic factor $\Phi_{(j)}$ taking account of the dynamic magnification of stresses and vibration effects in the structure, but unable to predict resonance effects from high speed trains. Generally the dynamic factor Φ_1 is taken as either Φ_2, according to the quality of track maintenance and determinant length L_Φ allowing these factors to be used for other structural members with different support conditions.

1.1.2 Dynamic Behaviour Under Traffic Loadings

This simplified procedure cannot be adopted at all times for higher speed, as the ratio of maximum stress due to a train crossing at greater speed to the maximum stress due to the same train journey very slowly is usually a function of several parameters. The susceptibility of appearance of resonance in a determined bridge deck depends thus not only on the value of the train loads, but also on their spatial distribution, the train speed, the rigidity of the convoy, the irregularities of the rail and the natural frequencies of the deck itself. Therefore, complex modelling not only of the bridge, but also of the train groups is used in order to properly analyse their dynamic interaction. It is first important to make sure that the wheel and rail contact is still maintained despite the oscillations of the structure and the train dynamic trajectory As a result, the arising vertical acceleration onto axles and the track twist due to the girder torsional movements have to be restricted. Secondly, it is also necessary to check that the girders dynamic oscillations do not cause a reduction in track stability or loss of track geometry. Safety and operational comfort aspects impose more strict limitations on the values of the acceleration. To prevent any discomfort when a train is crossing a bridge, passengers should not be subject to excessive levels of vertical acceleration. These accelerations are generated, on the one hand, by bridge oscillations and, on the other, by the damping from vehicle body suspensions. Depending on the level of comfort required, the vertical accelerations on vehicle bodies are limited to a value between 0.1, 0.13 or 0.2 g.

1.1.3 Serviceability Requirements

Safety and comfort are two major requirements to determine also the deformability limits of rail bridges. Therefore, new design bridges should not be made more flexible than existing ones. To check the permissible vertical deflection for speeds less than 200 km/h to minimise track maintenance and to avoid dynamic studies, $\alpha = 1.0$ shall be adopted. Even if $\alpha = 1.21$ or 1.1 was taken into consideration for *ULS*. This simplified rule for speeds less than 80 km/h, permissible values for deflections calculated under *LM 71,* multiplied by dynamic factor Φ, should not exceed *L*/600.

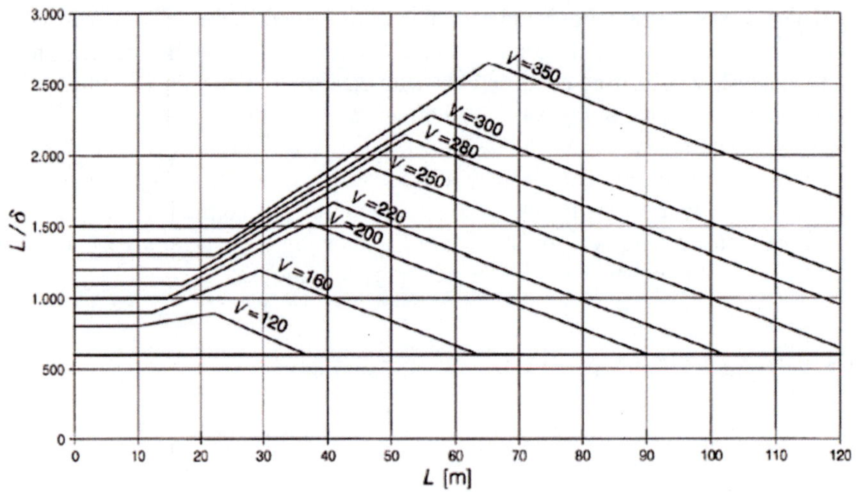

Fig. 1 Recommended permissible deflections

For greater speeds within the range $80 \leq v \leq 200$ km/h as it can be seen in Fig. 1 from EN 1991-2 [2], the permissible deflections δ_{stat} result from expression

$$d_{stat} \leq L/(15\,v - 600) \tag{1}$$

The upper limit *L*/2400 for 200 km/h is the permissible deflection which has taken during many years for designing bridges for high speed lines with satisfactory results. For speed superior to 200 km/h, the dynamic study is obligatory. Determined values should be minimally $\delta_{stat} \leq L/2400$. Such limitations should guarantee that the dynamic stresses due to actual trains at speeds smaller than or equal to 200 km/h remain smaller than the stresses calculated with the *LM 71* loading scheme, including the dynamic factor. At the very high speeds in excess of 200 km/h, this check has to be supplemented by dynamic calculations under actual traffic in order to cover any resonance or excessive vibration of the girder.

The supplementary limitations on the displacement of deck ends beyond bearings due to traction and braking are required for safety and comfort reasons of higher operational velocity, as indicated in Fig. 2. The relative longitudinal displacement δ_2 shall not exceed 5 mm for continuous welded rails without rail expansion devices or with a rail expansion device at one end of the deck and 30 mm for rail expansion devices at both ends (Fig. 2a). For vertical traffic actions produced by non-classified model *LM 71*, the longitudinal displacement δ_3 shall not exceed 8 mm when the combined behaviour of structure and track is taken into account (Fig. 2b). The vertical displacement of the upper surface of a deck relative to the adjacent construction or another deck abutment δ_4 due to variable actions shall not exceed 3 mm for a maximum line speed at the site of up to 160 km/h and only 2 mm over this value (Fig. 2c). These

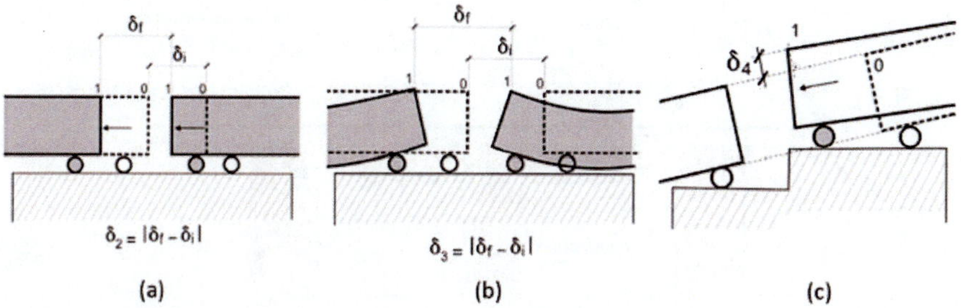

Fig. 2 Limiting values for the deformation of the structure

quite restricted displacement values are rather demanding in a practical design, even in the case only higher speed tracks.

The railway bridges, especially for higher operational velocity, must be carefully designed and constructed in a fatigue-resistant way, for having optimal life cycle costs and for reaching the intended design life of minimum 100 years. Accordingly, all important structural members shall be designed for fatigue. Constructional details have to be chosen and found to provide the maximum possible reference value of the fatigue strength $\Delta\sigma_c$, at least of detail categories 71.

1.1.4 Bridge and Track Interaction

When a track is continuous at least at one end of the bridge, the longitudinal forces generated by the track are distributed as a result of the interaction between track and structure. The longitudinal force components transmitted to each element depends on track resistance to longitudinal displacement in relation to the adjacent structure or substructure, and on the girder resistance to longitudinal displacement, hence on the stiffness of bearing devices, piers, foundations. The loading cases likely to generate additional horizontal forces are essentially thermal expansion, horizontal traction and braking loads. The additional forces will have to be withstood by the track. But the force components affecting the bridge will have to be taken into consideration for the design of the structure. Theoretically this is a serviceability limit state for the bridge and an ultimate limit state for the rail. The expansion length between the thermal fixed point and the end of the deck, number of spans and length of each span, position of fixed bearings and the thermal fixed point are basic parameters affecting the combined response of the structure and track. The principle of this problem solution was published by Esveld and Kok in the document [4]. A more complex answer is based on the standard EN 1991-2 [2]. For the determination of load effects in the combined track and structural system a model based upon Fig. 3 may be used. Practice with rail *UIC* 60 of the steel grade giving at least 900 N/mm^2 strength, placed on ballasted track with concrete sleepers and consolidated minimum

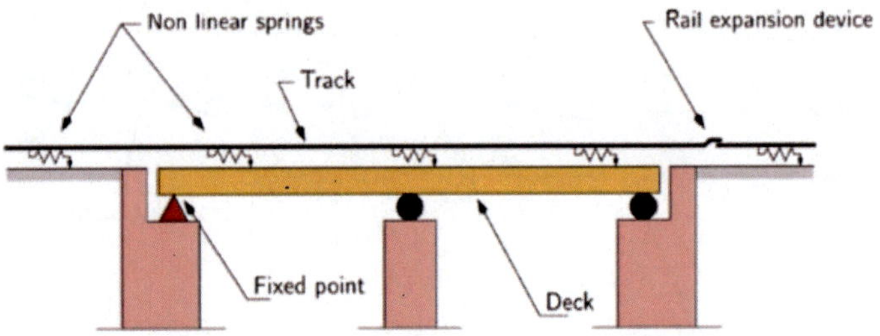

Fig. 3 Model of structure for interaction

300 mm deep ballast, the permissible additional stresses in continuous welded rail on the bridge due to interaction are usually 72 N/mm^2 in compression and 92 N/mm^2 in traction.

1.1.5 Approach Banks Construction

Bridge approaches can receive sudden impact loading due to a change in the vertical stiffness between the bridge deck and approach embankment leading to frequent settlement of the approach bank, as indicated in Fig. 4. This may create large acceleration at high speed leading to riding discomfort. Abutments provide the vertical support to the superstructure at the ends and connect the bridge with the approach track. The larger backfill volumes often result in more settlement of track approaches. Proper design of approach banks can include also provision of wider sleepers at approaches, an approach slab or providing tapering stone to simulate gradual rise in approach stiffness. Effective concept is similarly the use of multiple-layer strips or fibres to reinforce the fill material perpendicular to the wall face to minimise the large relative displacement between the abutment and its backfilled soil. Drainage shall be suitably provided for the abutment construction, consisting of penetrable materials, pipes and weep holes embedded in the backfill soil. The system should

Fig. 4 Approach banks arrangement

be properly designed to reduce the possible accumulation of hydrostatic pressure, to control erosion of the roadway embankment, and to reduce the possibility of soil liquefaction.

2 Types of Bridge Structures on Higher Speed Lines

2.1 Simply Supported Bridges for Small and Medium Spans

Simply supported bridges are the most commonly constructed superstructures on the retrofitted high speed lines. One important reason for the popularity of simply supported structural systems was that they could be replaced easily and rapidly without disturbing the substructure and adjacent structures.

Filler concrete beam deck bridges with encased rolled or welded *I* or *H*-shaped section are still the most typical and widely used type of bridge structure on the actually modernised network [5]. Additionally, some studies consider also untraditional shape of the pieces, particularly the upside-down *T*-sections. Especially for small overpasses, typical span lengths of these encased steel beams are from 8 to 16 m. The effectiveness follows from the favourable properties of basic materials, stiffness and strength of concrete in compression and great strength of steel in tension. Particularly from the viewpoint of more strict criteria limiting deformations and improving dynamic behaviour. Small structural depth represents another significant advantage of the filer-beam deck bridge. The current global analysis of the bridge structures is usually simplified to the calculation of single beam model. Besides space three dimensional finite analyses can be actually applied. A plate model of deck with steel beams approximated by ribs encased into the deck may provide the improved load redistribution resulting in more accurate internal forces and deflections.

An example of a successfully executed bridge composed of three simply supported spans 7.50 + 10.45 + 7.50 m long can be seen in Fig. 5. This railway overcrossing with composite concrete slab has incorporated eight hot rolled steel profiles *H 320B* in shorter spans and *H 450 M* sections in greater medium span with transverse

Fig. 5 Analyses modelling of railway crossing of encased steel beams

reinforcement made of steel bars. Thus, the achieved structural heights of the spans only 0.54 m and 0.60 m, respectively, resulted in the very slender superstructures. Vertical reactions of adjacent spans were transmitted via elastomer coupling bearings, and the transverse forces absorbed by additional bearings on top of abutments and pillars.

The low-maintenance characteristics of slab track concepts are being applied in the Slovak railway corridors, although still at a moderate volume. The slabs are obviously concreted on site. Increased service life, high lateral track resistance, no shaking of ballast particles are the further advantages of such structures. As an example, the ballastless track structure was recently built in the frame of the modernization, especially in the tunnel of "Turecký vrch" and at its adjacent bridge. The slab structure of RHEDA 2000® [6] type consists of rails 60E2, Vossloh 300-1 fastenings, track supports in the form of twin-block concrete sleepers and the monolithic reinforced concrete slab 300 mm thick of C35/45 concrete class, as shown in Fig. 6. Considering the resulting geometrical parameters of the structure, the investment costs and the need for reasonable maintenance interventions, the ballastless track seems to be also suitable for lines of higher speeds.

Steel solutions can be optimal answers in situations, where reducing the mass of the bridge becomes mandatory. Some examples are high piles viaducts or bridges in areas with low geotechnical capacity. In these cases, the conjunction of the lightness of the steel with the robustness of concrete in composite designs can reduce up to 2 or 2.5 times the total weight of the deck compared with alternative concrete designs, proportioning at the same time the necessary stiffness. But even presently, the most common solution takes still advantage of traditional ballasted track also in high speed lines. The choice of the *UIC 60* rails is recommended from technical and economic reasons. The selection of the type of fastening system depends essentially on the sleeper type used and on the stiffness of the granular layers which support the sleepers. The most common typology is the mono-block or modified bi-block

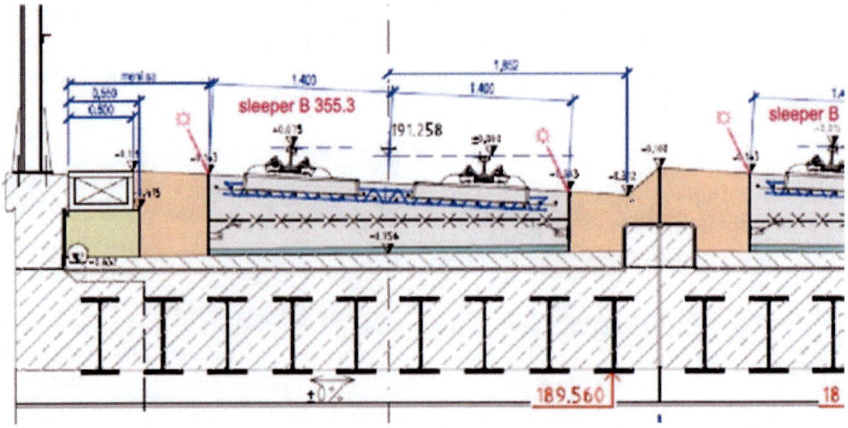

Fig. 6 Slab track at the 12 m span composite bridge with encased H 300 B steel profiles

sleepers and those elements are generally placed at a distance which can vary between 500 and 700 mm. The ballast layer not inferior 400 mm in depth is adopted in order to guarantee the capacity of spreading and conveniently transmitting the loads that are transmitted to the supporting structure without damaging the surface of a bridge. However the influence of the mass of the bridge is quite significant. Increasing simply the mass produces slightly unfavourably reduction of the critical resonant speed but also beneficial decrease of the maximum vertical acceleration. But simultaneous optimal increment on the mass and also the stiffness of the bridge may produce a decrease of the maximum deflections and accelerations, without affecting the critical speed.

For the aforementioned reasons, the steel bridge design is enabling them to gain access into the higher speed lines, demonstrating their competitiveness and reliability. In order to give an idea of the adaptation of the main structural characteristics of ballasted steel bridges to the above requirements, a simple one span 30 m long plate girders superstructure as another structural type used for crossing river is represented in Fig. 7. Two side steel plate girders fabricated from web plate 14 mm thick had to be reinforced by vertical stiffeners along the length and joined together with lower flanges from flat steel 700 mm large and 50 mm thick. Floor beams 525 mm high connected to girders from side to side and longitudinal stringers 250 mm high running between these cross beams could create an orthotropic deck. This steel bed holds ballasted deck with standard track on the ballast. Bottom and cross bracings also contribute to stability. Figure 7 shows also the developed finite transformation model.

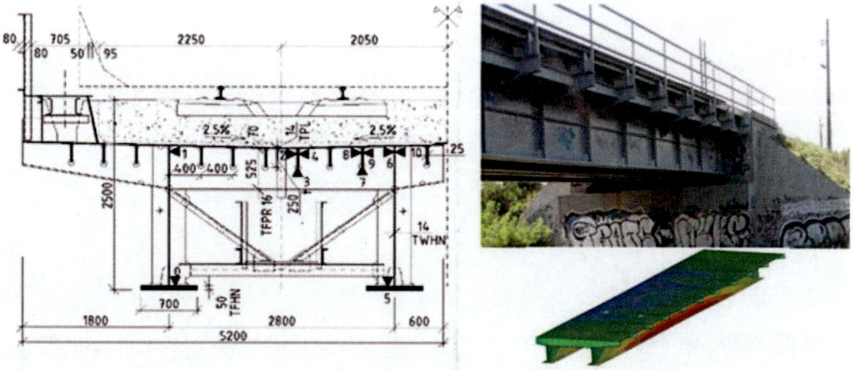

Fig. 7 Ballasted single span 30 m long bridge and its finite numerical model

2.2 Bridge Types for Larger Spans

Trusses can require a very small amount of material for construction and result in a cost-effective bridge form, thanks to forces distribution across almost entire structure span using a diagonal mesh of triangle-shaped. Moreover, the single span Warren truss without verticals in the parallel chord shape is also aesthetically pleasing truss configuration as exemplified such a bridge arrangement shown in Fig. 8. Two single spans railway bridge has been rebuilt on existing route to structural system having two identical spans of length 80 m. The low-maintenance renewed superstructure is 6.65 m wide, set by clearance space required for passage of contemporary higher speed rail traffic. The trough truss bridges with ballasted deck consist of two truss main girders having 10.0 m height on sliding elastomeric pot bearings at the interior pier. The both chords and diagonal are 0.5 m wide with variable plate thickness adapting cross sections to bending moments along a span. The fully welded box shape sections are used for compression members, as they have lower slenderness. In the case of tension members, the choice of H shaped sections was encouraged by relatively clear advantage of easier connection to gusset plates, without the stability design requirements needed. Also this system has led to a significant simplification in truss detailing, because sway frames are typically omitted in this form of truss, except for portals.

The orthotropic deck for ballasted track bed has been used as part of the bottom chord on this bridge. This would lead toward loading the chord at other than the panel points. Thus, the lower chord members directly support the deck over their full length, not simply at the panel points. Merging chords and deck has also permitted eliminating more joints within the deck system, as favourable improvement. The available computer capabilities allow modelling of redundancy issues, necessary considered as compared to traditional solutions. Appropriateness of the model was confirmed by a comparison of values produced by numerical analysis and those

Fig. 8 Parallel chord Warren truss option for two single spans, each 80 m long

obtained from measurement. In the static loading test, a heavy emergency railway crane EDK 750 was used as live-load in specified configurations.

Arch bridges are generally considered to be better looking than truss systems as far as aesthetics in challenging, mainly city environment. As next advantage, the single span tied arch reacts on the supporting substructure as if it were a simply supported beam. Internally, however, the system is indeterminate with the behaviour being dependent on the ratio of the tie stiffness to the other structural element stiffness. Thus, as the next case study, the parallel superstructures are both single-arch bridges over the river with a span of 112.0 m. The arch rise is 18.0 m, resulting in an arch-to-span ratio of 1/6. The main girders, axially 7.1 m spaced, connecting the ends of the arch and participating with the deck are welded plate girder *I* sections. They have the slender web 16 mm thick and construction depth 2.8 m. The upper flange 650 mm and lower one 800 mm large, both are of the constant thickness of 30 mm. The arch rib very nearly in pure compression was made up of plates 25 mm thick in the shape of a rectangular box 700 × 600 mm. The vertical suspenders are located every 8.0 m for long spans and consist of welded*I*section 220 mm high. The bracing include *K* type arrangement shown in Fig. 9. The orthotropic steel deck of ballasted track participates with the main longitudinal support system [7].

In response to the stresses induced during a train passage, ballast experience plastic settlement which can reach unacceptable magnitudes, especially in the zone of movable bearings, as indicated in Fig. 10. Ballast improvement techniques can use geogrids as one of solutions to cope with increasing train speed, load and frequency.

Fig. 9 Parallel arch tie-girder bridges of bow-string system

Fig. 10 Ballast irregularities at the bridge ends

But, the fibre reinforcement in ballast is actually preferred, as new method of reinforcing ballast and maintaining the capability of typical tamping maintenance operations incompatible with geogrids. The addition of propylene fibres from 11 to 47 mm in length and 0.12 mm in diameter to sand and scaled ballast can minimise the voids in the mixture as fibre content can increase the compactness.

Continuous box girder as an occasionally preferred option in design of higher speed bridges, represent another structural type used, especially for crossing of longer obstacles. Compared to simply supported systems, this bridge type not only exhibits greater slenderness but also reveals lower dynamic responses. The slenderness ratio, as span length to cross-section height ranges from 11 to 18. Continuous box girders can be designed as having either uniform or variable depth.

As an example, the continuous box girder bridges in Fig. 11 are 341.0 m long and consist of seven spans, which have the following span configurations $40.5 + 3 \times 52.0 + 53.0 + 53.4 + 40.5$ m and $40.5 + 3 \times 52.0 + 2 \times 50.8 + 40.5$ m, respectively. The superstructure box cross-section height varies between 3.12 and 4.62 m, which translate to a slenderness of 11.6 and 17.2. A simply supported bridge at this location would require a pier cross-section width of 4 m, as opposed to the actual width of 2.7 m.

As shown in Fig. 12, Rheda® ballastless track system was selected. At first, the cross-shaped base slabs were directly concreted on the protective layer of bridge deck.

Fig. 11 Seven spans post–tensioned box–girder bridge

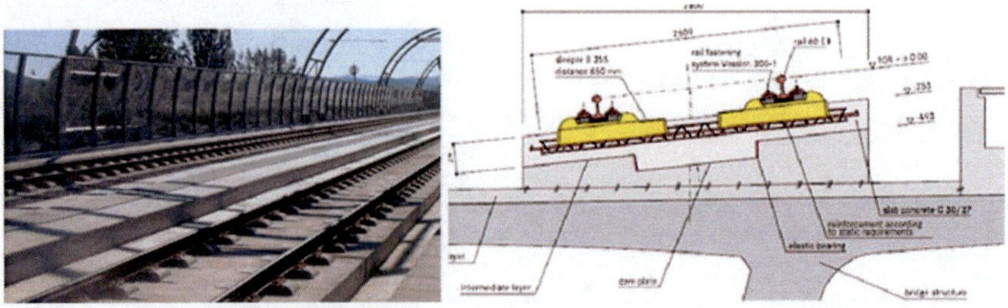

Fig. 12 Ballastless track assembly

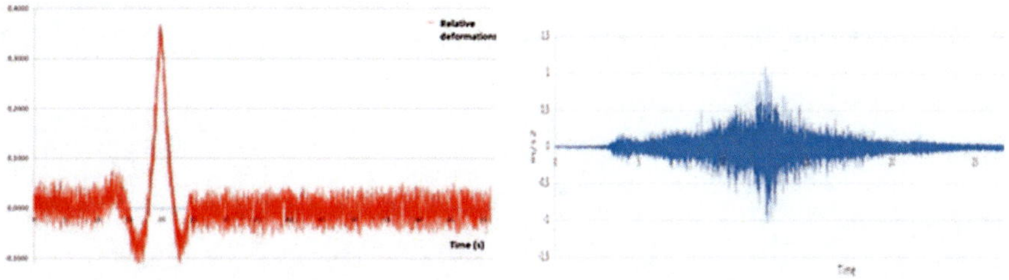

Fig. 13 Relative deformations and vertical acceleration at the first critical velocity of 50 km/h

After that, the segmented concrete slabs of C30/37 grade with nose as the reinforced concrete stoppers 800 × 800 mm and 150 mm high were installed in pre-designated zones. Transversal and longitudinal positions of them on the bridge deck were ensured by means of these elements. The elastomeric bearings stuck on edges of the stoppers can tolerate movements in the longitudinal direction. Thus, the overloading of the connexion, as a result of expansion, temperature or deck concrete rheology can be avoided. This ballastless track on this rather long bridge was preferred due to a range of advantages, when compared to the conventional ballasted track. The innovative type of railway track stands out mainly for its performance, leading to substantial reductions in maintenance costs and also in maintenance work like tamping, ballast cleaning or track lining. Despite the high initial construction costs, these expenses may be recompensed over the service life of the track, creating a more economical and competitive solution when assessing a broader dimension of time. Additionally, the problem with drag forces at ballast due to the passage of high speed trains is no more a reason for concern (Fig. 13).

The mid-span deflection in the sixth and major mid-span due to trains crossing of the bridge at a speed of 200 km/h obtained from calculations amounts less than L/4500. The extreme vertical acceleration identified in the field investigations was less than 0.53 m/s^2 and that was at the critical resonance speed of only around 50 km/h.

3 Concluding Remarks

Conventional steel bridges with open-grid deck flooring are replaced by through ballasted deck providing more stable and easily maintained level base for the rails, also reducing vibration and noise during traffic. Following a great amount of investigation and analysis, ballastless slab track systems on bridges were developed, along with new designs. In the paper, quite special bridge superstructures of this type are presented. The computer modelling using more sophisticated analyses is discussed and experimental assessment given.

Acknowledgements The paper presents results of the research activities supported partly by the Slovak Science Grant Agency; grant No. 1/0343/18.

References

1. https://ec.europa.eu/transport/themes/infrastructure/ten-t_en
2. EN 1991-2 Eurocode 1 (2006) Actions on structures—part 2: traffic loads on bridges. CEN
3. Gandil J, Tschumi MA, Delorme F, Voignier P (1996) Railway traffic actions and combinations with other variable actions. IABSE Reports, Zürich
4. Esveld C, Kok AWM (2013) Interaction between moving vehicles and railway track at high speed. In: Proceedings of the first international workshop. Issues & Challenges, UIC Paris Ballast
5. Bujnak J, Odrobinak J, Kvocak V, Kozlejova V (2019) Composite steel and concrete structures. Some aspects of appropriate design. University of Zilina, Monograph, p 102
6. Rheda® ballastless track system. https://www.railone.com/
7. Bujnak J, Bujnakova P (2020) Bridges. Conception and construction. University of Zilina, Handbook, p 186

Peculiarities of Reconstruction Work as a Result of Possible Terrorist Attacks on Out-of-Class Bridges

O. Shymanovskyi, V. Shalinskyi, and W. Baran

Abstract The paper sets out theoretical foundations and applicative recommendations regarding reconstruction work on out-of-class bridges as a result of possible terrorist attacks. It is determined that the traditional building design system under variable loads was developed on the basis of a thorough study of repetitive phenomena (for example, loads arising at regular intervals, ranging from moderate to a strong degree of intensity), which can be predicted through a statistical approach. In contrast, cases of explosions or intentional strikes during terrorist attacks are characterized as very low-probability loads, although they lead to the most catastrophic consequences. Close interaction between explosion loads and typical construction loads is one of the important factors that has only recently begun to be considered in practical design activities. And so, in many cases, it is now necessary to take into account all the loads, however rare or unbelievable, such as those caused by the attacks of criminal elements on buildings and engineering structures in view of terrorists' intentions to use destruction of facilities to achieve their objectives. In this sense, it is very important to recognize early the emergency situations and ensure the security of objects of possible terrorist attacks, which most often belong to objects of strategic vital importance (government facilities, high-risk facilities, unified transport infrastructure objects, especially important electric power industry facilities and places of mass stay of people etc.), and take preventive measures to ensure the operation of the main load-bearing construction systems and structural members which are sufficient to counteract the development of such situations. The two most important bridges in Kyiv across the Dnieper River are considered as objects of study: the Paton Bridge, 1542.2 m long (the world's first all-welded bridge with the girder system of such a length) and the Darnytsky Rail-Road Bridge, 1061.6 m long. The paper summarizes the description of structural solutions for bridges. Possible variants of terrorist acts

O. Shymanovskyi (✉) · V. Shalinskyi
V. Shymanovskyi Ukrainian Institute of Steel Construction, 2/1, V. Shymanovskyi str., Kyiv 02125, Ukraine
e-mail: ashim@urdisc.com.ua

W. Baran
Faculty of Civil Engineering and Architecture, Opole University of Technology, ul. Katowicka 48, 45-061 Opole, Poland

Z. Zembaty et al. (eds.), *Environmental Challenges in Civil Engineering*, Lecture Notes in Civil Engineering 122,
https://doi.org/10.1007/978-3-030-63879-5_2

on bridges and methods for eliminating their consequences are examined, taking into account the actual state of their structural members determined by the results of special surveys. The method is described for consideration of detected defects of bridge structures in design models of bridge constructions and of the blast wave impact, in order to assess the influence of these factors on the stress–strain state of bridge structures. The most probable variants of bridge structures destruction are sequentially considered, the most likely schemes of bridge structures destruction are determined and design solutions for replacing damaged sections of the bridge superstructure are proposed.

Keywords Bridge · Reconstruction · Destruction · Attack

1 Introduction

It is clear that the probability of a terrorist threat is directly related foremost to objects of strategic importance, in particular to out-of-class bridges, the proper functioning of which is undoubtedly the key to stable transport connection. Therefore, we will consider the variants of acceptable destruction of some out-of-class bridges as a result of terrorist attacks and methods of their fast reconstruction.

1.1 Forecast Consequences of Terrorist Attacks and Methods for Their Elimination

It seems indisputable that the main and most acceptable method to comprehensively avoid the destructive effect of the blast wave on buildings and engineering structures is to prevent the terrorist act per se. An assessment of such threats and applied security measures to reduce the corresponding risks are described in detail in articles [1, 2]. Otherwise, that is, in view of a terrorist attack that is likely to occur, in order to minimize its consequences, it is necessary to consider beforehand not only potential scenarios for destruction of the object, but also methods for its fast reconstruction. And therefore, the following are conceivable variants for the bridge structures' destruction as a result of the terrorist attack and design solutions for the necessary reconstruction work on the example of two out-of-class bridges in the city of Kiev across the Dnieper River: the Paton Bridge and the road part of the Darnytsky Rail-Road Bridge crossing.

2 Paton Bridge Across Dnieper River

Brief description of bridge structures. The bridge was built in 1953 (Fig. 1).

Fig. 1 Panoramic (**a**) and general (**b**, **c**) view of Paton bridge

The world's first all-welded girder bridge of 1543.0 m long (Fig. 2a). The carriageway width between the curbs is 21.0 m and the width of pedestrian paths is 3.0 m. The bridge was designed according to the standards of 1948 to allow four highway convoys with a width of 3.375 m each (excluding stop ways) and two tram traffic lanes (tram tracks are now demounted) with a total tramway width of 7.5 m (Fig. 2b).

The bridge spans are: in the bottom land part—58.0 m, in the channel span part 87.0 m and near the banks—17.1 m. The bridge superstructures are accepted as all-welded continuous girder system consisting of four main girders, which are combined in cross-sections with bracing made of angles. The breakdown into continuous spans is taken as follows: $(4 \times 58) + (4 \times 58) + (58 + 4 \times 87 + 58) + + (5 \times 58) + (5 \times 58) = 1508$ m.

The main girders have a web with a height of 3600 mm (which is enlarged to 6100 mm with straight-line haunches on the channel span piers) and are made of M16C carbon steel, which characteristics correspond to steel ВСт3сп. Transverse and longitudinal bracing are accepted in form of through trusses made of angle section members.

Probable variants of terrorist attacks on the Paton Bridge and methods for eliminating their consequences. First of all, it should be pointed out that during the last special examinations of the Paton Bridge a number of defects were revealed in bridge structures [3–6], which not only negatively affect the durability of the latter, but in some cases even lead to overall bearing capacity decrease. Therefore relevant studies were conducted, in order to assess the influence of these factors on the stress–strain state of bridge structures, as well as to develop constructive and technological recommendations for eliminating the consequences of terroristic acts. For this, the existing defects were implemented in the bridge structure finite-element model, which was created by selecting all the initial data from the working drawings when performing the primary strength and dynamic calculations of the Paton Bridge using the geometry design values. However, due to significant number of identified defects, prior to their introduction into the finite element model of the bridge, their

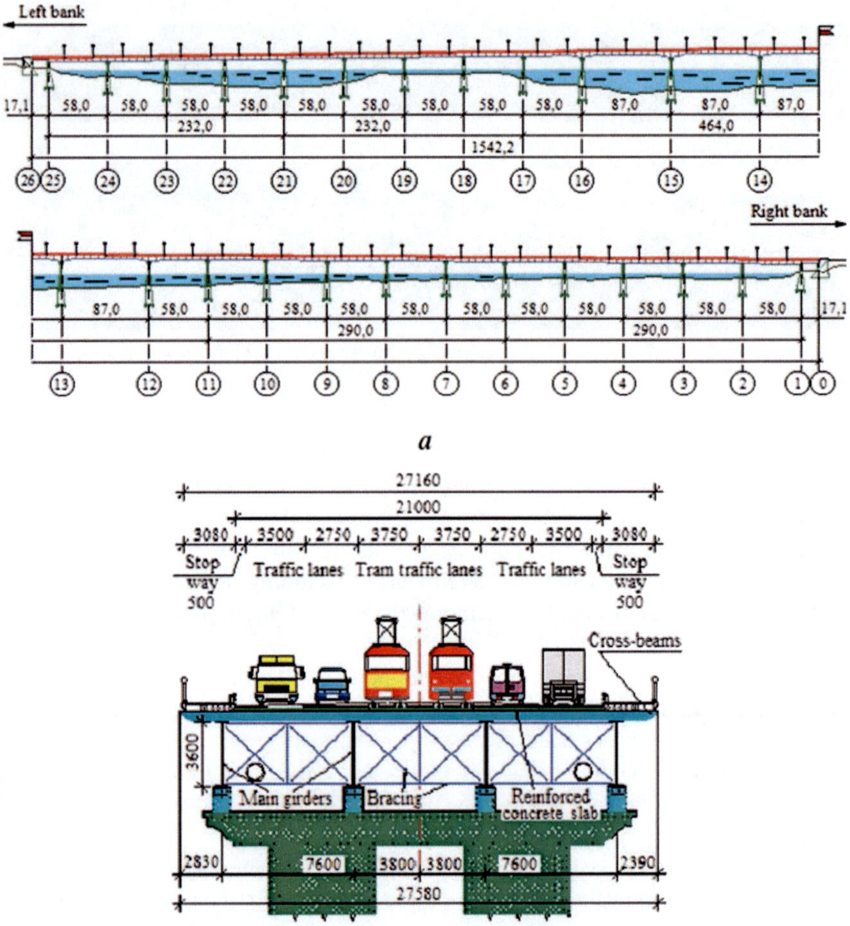

Fig. 2 Longitudinal (**a**) and transverse (**b**) sections of Paton bridge

systematization and a certain typical averaging were performed. As a result, it was accepted that, depending on the defect types, they can be reproduced in the model in two ways: either by direct extracting the corresponding finite elements in the places where the holes, openings and slots observed during the inspection were located, or by adjusting the thickness of the bridge structure elements depending on the depth of their corrosion damage while maintaining other geometry and stiffness parameters.

Numerical studies also contained detailed analysis of accepted design decisions of the Paton Bridge across the Dnieper River and the blast wave effect on steel constructions with the use of relevant methods elucidated in [1, 2]. As a result of these calculations, it was found that when a terroristic attack is carried out within the bridge framework in the case of using an explosive device with an estimated weight of 160–200 kg (depending on the place of the attack), this framework will be completely destroyed and cannot be renovated. However, the adjacent bridge framework will generally survive, and therefore it is quite possible to operate it further after necessary repairs have been completed. In this context, we draw attention to the obvious fact that the indicated amount of explosives can be delivered onto the bridge exclusively by means of a vehicle, because a person alone is not able to bring such an amount to the bridge.

So, during the numerical studies of the Paton Bridge, taking into account the existing defects in developed technical solutions for reconstruction work on the bridge, four most likely variants were considered for the bridge structure destruction, which are schematically shown in Figs. 3, 4, 5 and 6.

- variant—the bridge superstructure of 58.0 m span is destroyed over the entire width of the bridge in the piers' axes 6–7 within the end span of continuous five-span girder (Fig. 3);
- variant 2—pier No. 9 of continuous five-span superstructure is destroyed, and adjacent spans of 58.0 m are collapsed (Fig. 4);
- variant 3—the bridge superstructure of 87.0 m span is destroyed over the entire width of the bridge within axes 12–13 in the second span of continuous six-span girder (Fig. 5);
- variant 4—pier No. 15 of continuous six-span superstructure is completely destroyed to the edge of the foundation, and adjacent spans of 87.0 m are collapsed (Fig. 6).

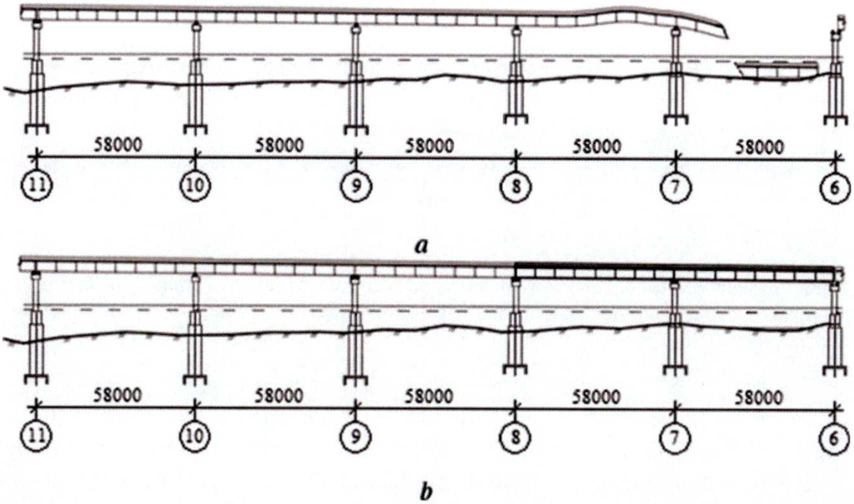

Fig. 3 Schematic structural diagrams of destruction and reconstruction of bridge structures in case of probable destruction variant 1: **a** superstructure after destruction of span 6–7; **b** superstructure after reconstruction

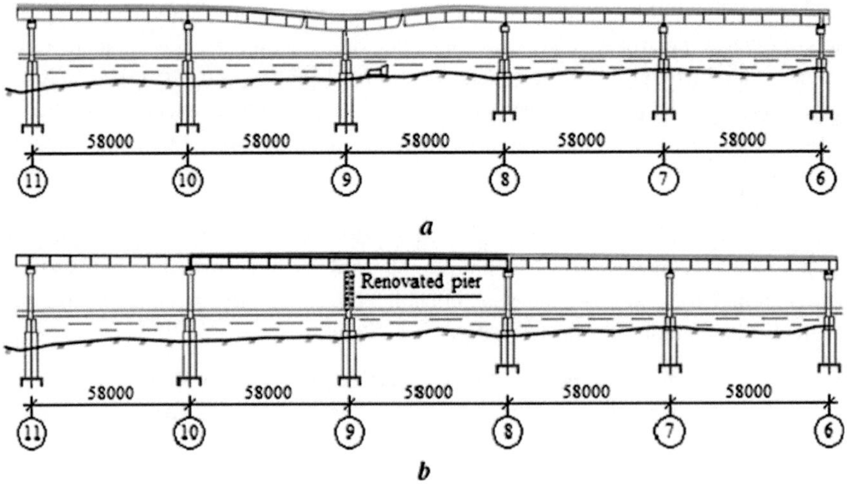

Fig. 4 Schematic structural diagrams of destruction and reconstruction of bridge structures in case of probable destruction variant 2: **a** superstructure after destruction of pier no. 9; **b** superstructure after reconstruction

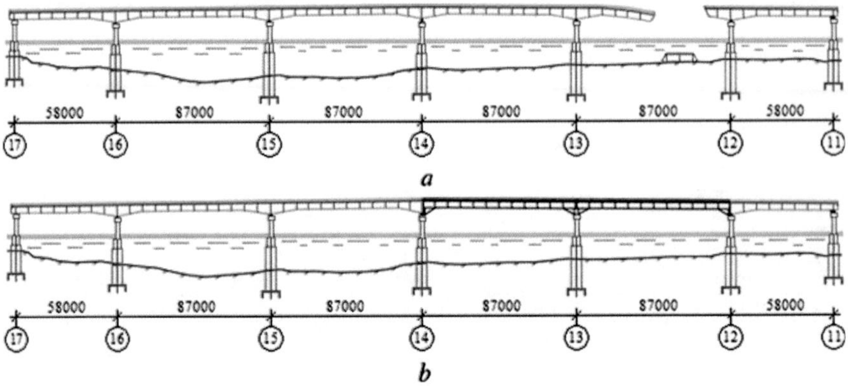

Fig. 5 Schematic structural diagrams of destruction and reconstruction of bridge structures in case of probable destruction variant 3: **a** superstructure after destruction of span 12–13; **b** superstructure after reconstruction

As for the essence of design solutions for reconstruction work, the following should be noted. Depending on the variants considered for destruction of bridge structures, technical recommendations were formulated for renovation work on each variant separately, based on the results of calculations and developed basic design solutions for superstructure and supports of the Paton Bridge. In particular, for all four variants considered for the bridge structures' destruction, superstructures which have collapsed or have permanent deformations of such magnitude that make them unsuitable for further operation are supposed to be replaced with new structures

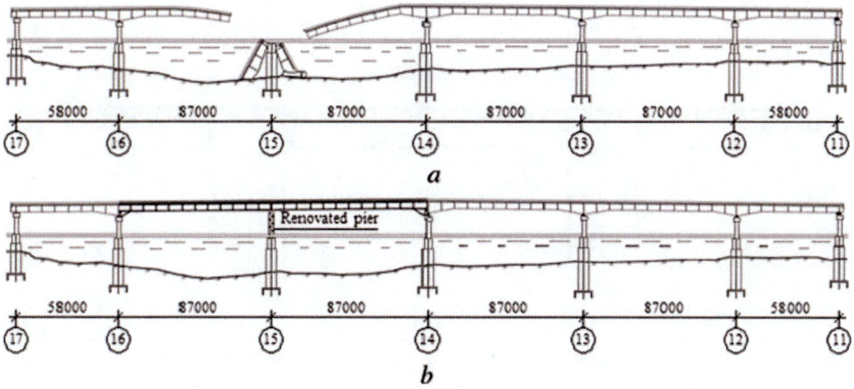

Fig. 6 Schematic structural diagrams of destruction and reconstruction of bridge structures in case of probable destruction variant 4: **a** superstructure after destruction of pier no. 15; **b** superstructure after reconstruction

using methods, conceptual sketches of which are reproduced in Figs. 3, 4, 5 and 6 using black to indicate reconstructed bridge decks. And the adjacent spans of the multi-span continuous girder, which remained undamaged and are shown in Figs. 3, 4, 5 and 6 in gray, as a result of destruction of one span, they turn into either simple girders or into two-span or four-span structures instead of five-span or six-span ones. Therefore, considering such significant transformation of the initial design layout, the forces in structural members of the adjacent intact girder spans are increasing, and the spans themselves according to the results of checking calculations will require some reinforcement.

It is proposed to replace completely destroyed spans with newly manufactured three types of similar inventory constructions of steel box girders and the carriageway in form of steel orthotropic deck, in the first of which the distance between two box girder axes is 9500 mm, in the second one—11,900 mm, and in the third one it is planned to install three box girders at a distance of 7300 mm between them (Fig. 7). It is interesting here that the indicated distances are caused by only one objective criterion—the layout of girders on the support between the bearing structures of undamaged sections of the spans. The height of the girders for 87.0 m span is taken 3600 mm, and for 56.0 m span—3100 mm. In addition, it should be noted that according to the results of comparing the three proposed types of inventory constructions, the second variant, in which two main girders are installed at a center distance of 11,900 mm, turned out to be the most economic and least labor-intensive for manufacturing and installation.

As for the design solution of proposed superstructure unit, it consists of seven elements of the same type, suitable in size for railway transportation: two units of box girders and five plates of the orthotropic deck. In this case, the plates are only two makes—the intermediate plate and the end plate. The main girders of the span are box-shaped, of rectangular cross-section, 2300 mm wide and 3100 and 3600 mm high, depending on the span. The lateral stiffness of the superstructure is

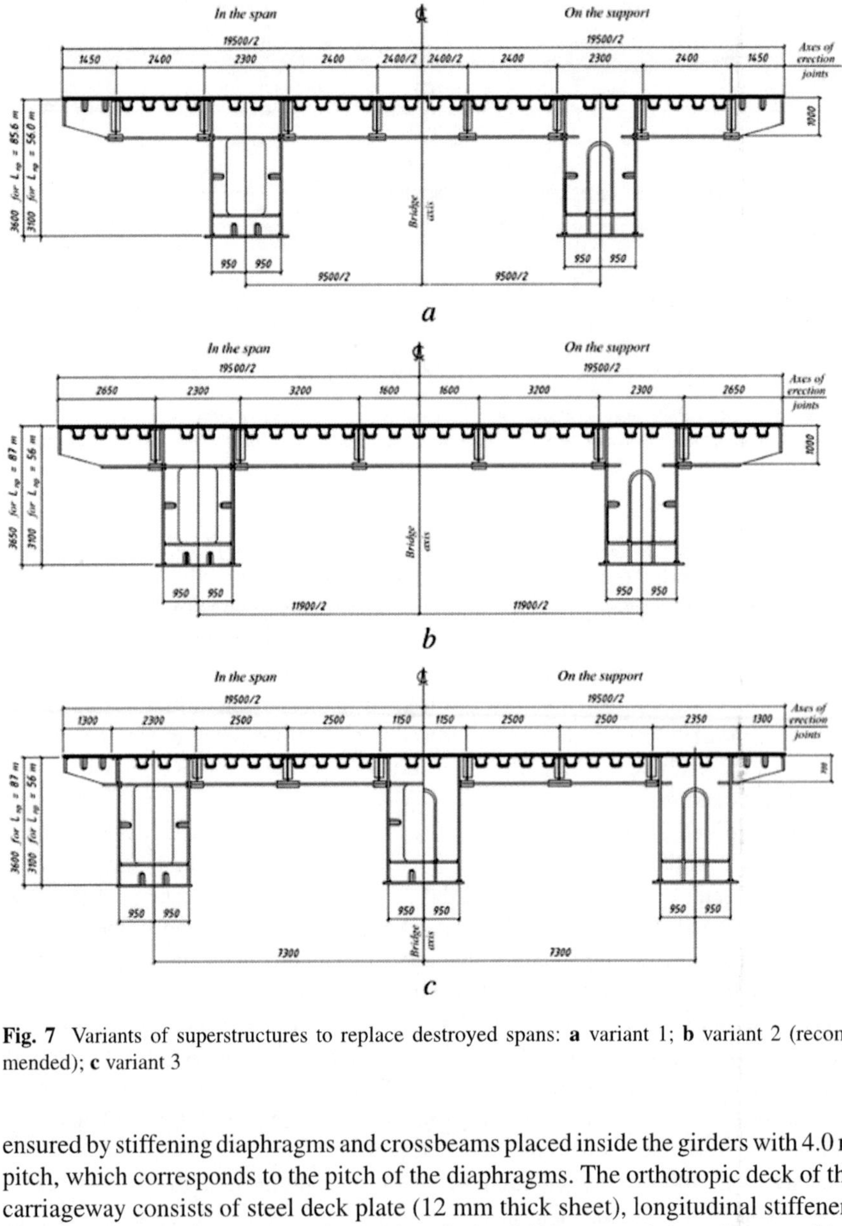

Fig. 7 Variants of superstructures to replace destroyed spans: **a** variant 1; **b** variant 2 (recommended); **c** variant 3

ensured by stiffening diaphragms and crossbeams placed inside the girders with 4.0 m pitch, which corresponds to the pitch of the diaphragms. The orthotropic deck of the carriageway consists of steel deck plate (12 mm thick sheet), longitudinal stiffeners of trapezoidal closed cross-section with a pitch of 600 mm and cross-beams with a pitch of 4000 mm.

As for the design solution of proposed superstructure unit, it consists of seven elements of the same type, suitable in size for railway transportation: two units of box girders and five plates of the orthotropic deck. In this case, the plates are only two makes—the intermediate plate and the end plate. The main girders of the span are box-shaped, of rectangular cross-section, 2300 mm wide and 3100 and 3600 mm high, depending on the span. The lateral stiffness of the superstructure is ensured by stiffening diaphragms and crossbeams placed inside the girders with 4.0 m pitch, which corresponds to the pitch of the diaphragms. The orthotropic deck of the carriageway consists of steel deck plate (12 mm thick sheet), longitudinal stiffeners of trapezoidal closed cross-section with a pitch of 600 mm and cross-beams with a pitch of 4000 mm.

And reinforcement of the chords of adjacent continuous superstructure spans (as considered above) is executed by welding additional plates to the chords of the main girders.

In case of destruction of the piers, their reconstruction is supposed to be executed of steel lattice structures consisting of I-sections and angle sections. Proposed design solution for lattice structures is shown in Fig. 8.

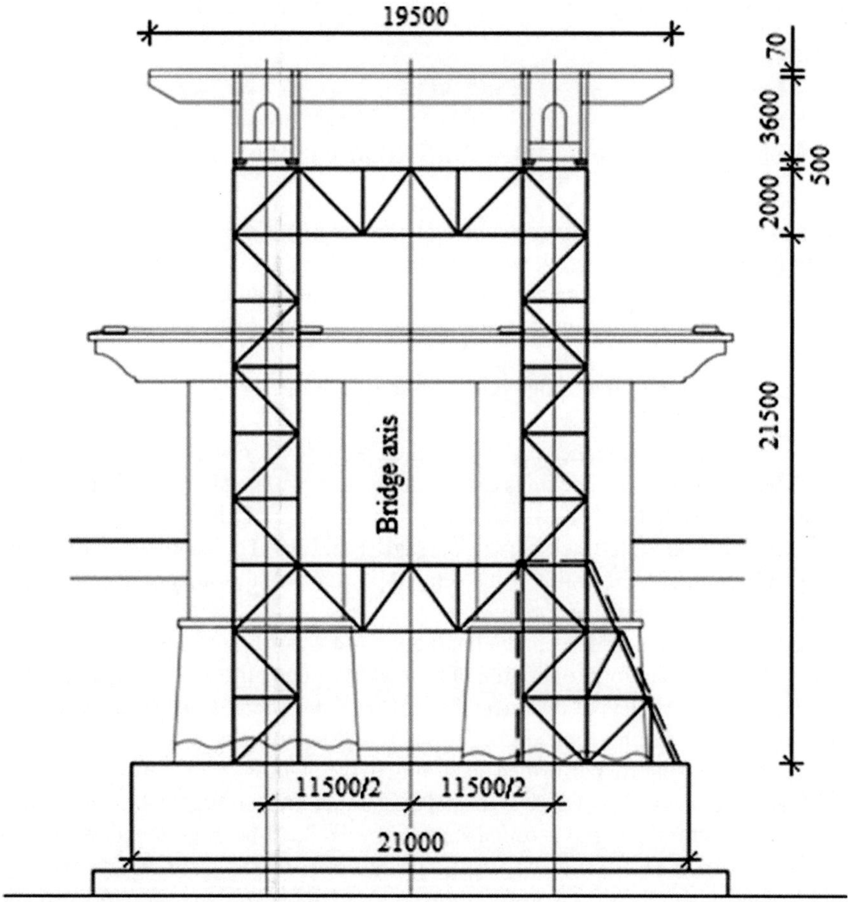

Fig. 8 Renovation of destroyed pier

Considerations on arrangement of work. The major premise for development of these technical solutions on reconstruction work implementation, depending on the destruction scheme under consideration, was to carry out the works as fast as possible, with minimal labor costs, with high manufacturability and process-ability for installation of structures. Therefore, in order to achieve this goal during reconstructive work, the use of monolithic reinforced concrete was almost completely excluded. The metal structures of supports and spans, which replace the destroyed ones, are recommended to be made beforehand at the metalwork plants and pre-assembled depending on the crane equipment so that they can be delivered to the installation site without hindrance. To do this, new spans are designed lightweight with a carriageway in form of steel orthotropic deck, and supports—metal lattice ones made of extensional units. Moreover, an additional significant advantage of the proposed design solutions is that in order to pre-assemble the spans, the Dnieper River embankment can be used, where convenient berthing facilities are for mooring and delivery of landing pontoons, barges and other floating crafts to the installation site.

3 Road Part of Darnytsky Rail-Road Bridge Crossing

Brief description of bridge structures. The Darnytsky Rail-Road Bridge crossing over the Dnieper River (Fig. 9) was built in 2004–2011 in order to improve transport links in Kiev, the stress state of which was caused not only by increased rail traffic at the Kiev railway junction, but also by excessive workload of existing city road bridges.

The total length of the bridge is 1061.6 m (Fig. 10a). The design capacity of the bridge road part is 35,000 vehicles per day. According to the current classification of city streets and roads, the road part of the Darnytsky Rail-Road Bridge crossing, which has two driveways, belongs to the category of regulated traffic main roads with three lanes in each driveway with a lane width of 3.75 m. Driveways are located on both sides of the double-track railway passage (Fig. 10b). The width of the roadways makes 13.25 m, and the width of the service passageways located on the outside of the driveways is 0.75 m (Fig. 11).

Road spans of the bridge channel part are made according to the layout $56.5 + 3 \times 111.6 + 56.7$ m in form of continuous metal five-span girders. The superstructure in cross-section consists of two metal boxes of constant height, connected with each other by the metal orthotropic deck. The height of the boxes is 3.24 m. On the flood-plain part, the spans are performed according to the layouts: $55.0 + 55.8 + 2 \times 56.1 + 55.4$ and $55.4 + 4 \times 56.1 + 55.4$ m, and their structure is similar to the channel one. The superstructure material is rolled steel for bridge load-bearing members made of steel grades 10ХСНД-2 and 15ХСНД-2 according to GOST 6713–91, the bearings and expansion joints are taken of company MAURER (Germany). Anticorrosive protection of span structures is to be carried out with materials of company JOTUN (Great Britain). Erection joints of metal structures are taken as welded and bolted

a

b

Fig. 9 Day-time (**a**) and night-time (**b**) overall view of the Darnytsky rail-road bridge crossing over the Dnieper River in Kiev

ones with high-strength friction bolts of 22 mm diameter, the sealing of which is to be made using Sikaflex silicone sealant.

As for the design solution of the bridge road part, it consists of an orthotropic deck, welded cross-beams of double-T shape of equal height between the main girders and of variable height outside, and of the main box-section girders. The pitch of the cross-beams is 3.1 m, which are fastened to the main girders using top cover plates with high-strength bolts in one level. On the bridge, on both sides of the driveway, metal overhead contact-line supports are also provided with a pitch of ≈ 21.7 m.

The guardrails «Bridge guardrail H4B» are taken of company SGGT (Germany) with a pitch of 1.1, 1.14 m high (at service passage) and 1.54 m (railway side), it is made of galvanized steel. Guardrail posts are fastened to the superstructure with M20 bolts. The steel grade of the guardrail parts is Ст3пс5 and Ст3сп5 according to GOST 14637-89.

Railings 1.1 m high are located on the service passage. The pitch of the posts of $80 \times 60 \times 7$ mm section makes 1.5 m, the handrail dimensions are $100 \times 70 \times 5$ mm, the lower and upper stringers are $50 \times 30 \times 3$ mm, and the filling is 20×20

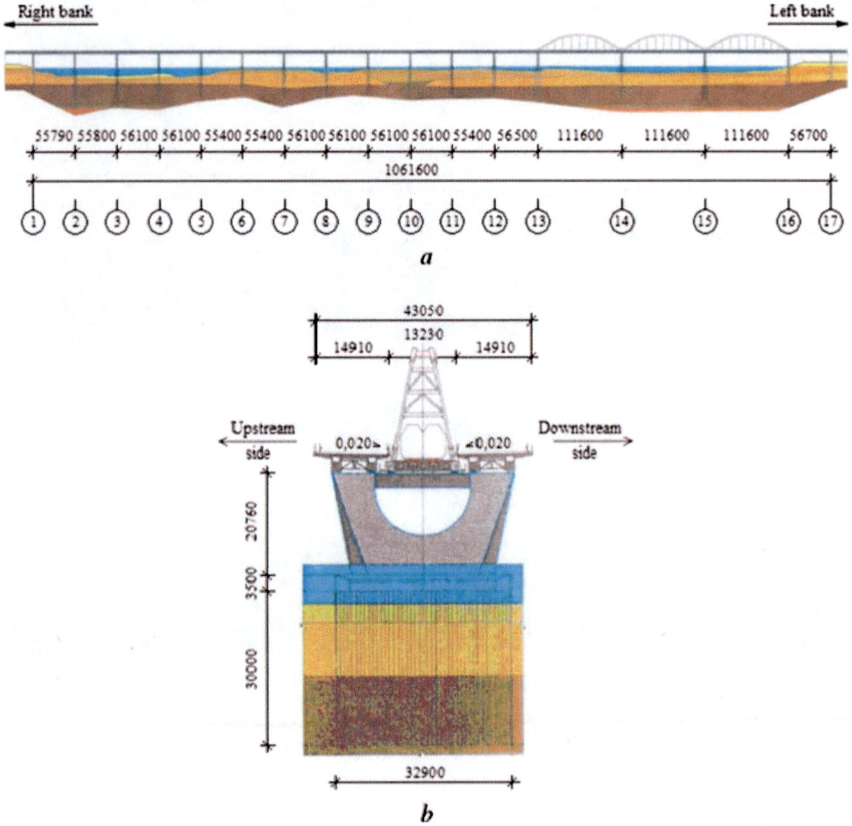

Fig. 10 Facade (**a**) and cross-section (**b**) of the Darnytsky rail-road bridge crossing

× 3 mm (150 mm pitch). All joints are welded, and expansion joints of the railing are provided over the superstructure movement joints. Structural material is carbon steel of Ст5пс grade according to GOST 14637-89. Paint system is to be made using Hardtop XP.

The pavement surface of the driveway and the service passage is of two layers of Gussasphalt hot-laid asphalt concrete, 80 mm thick (two layers of 40 mm each). The protective adhesive layer is waterproof coating based on Eliminator methyl methacrylate resin (four layers totally), 3.2 mm thick. As for the drainage system, we note that from the roadway it occurs due to a transverse slope towards the railway to the drainage pipe system (Fig. 10b).

Probable variants of terrorist attacks on the road part of the Darnytsky Rail-Road Bridge and methods for eliminating their consequences. First of all, it should be noted that the general methodology of numerical studies to identify variants of terrorist acts on the bridge crossing and methods for eliminating their consequences, in this case, is based on the approaches and principles highlighted above when considering the

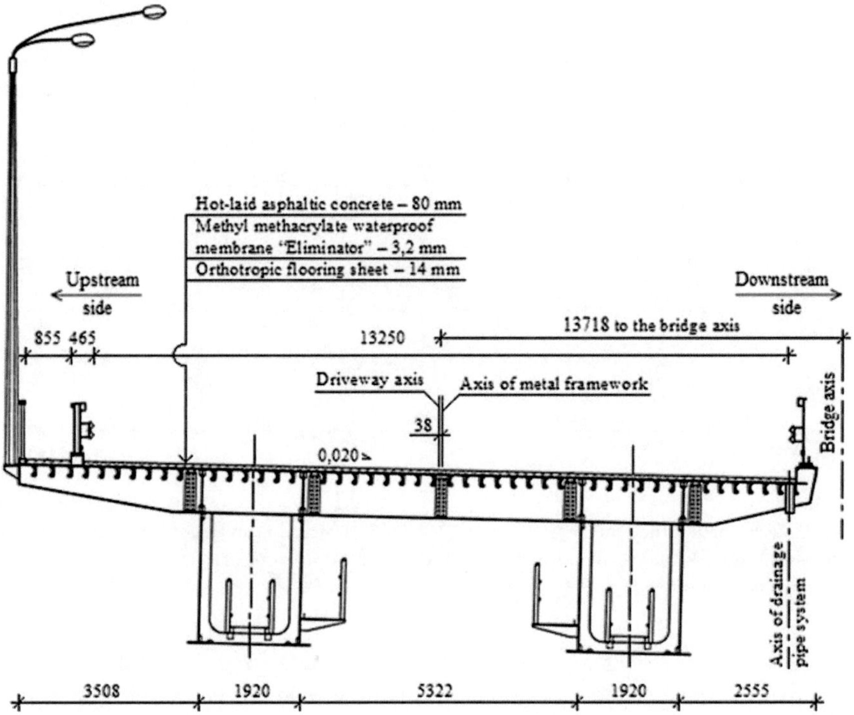

Fig. 11 Road cross-section of the Darnytsky rail-road bridge crossing on upstream side

Paton Bridge. And to detail the content of what has been said with such more specific words: the finite-element model of the bridge structures was built using initial design decisions and taking into account the results of special surveys conducted in 2018 on the road part of the Darnytsky Rail-Road Bridge crossing [7].

Taking into account previous considerations and the results of numerous studies, it was found that the place of a terrorist attack at the bridge is most likely within its continuous five-span structure, a schematic illustration of destruction through it is shown in Fig. 12a. And elimination of the consequences of this destruction is to undertake reconstruction work through implementation of three consecutive steps. And at the first of them, the remains of the destroyed five-span structure in the span, where the attack occurred, are to be dismantled. At the second step, two two-span structures remaining in place of the five-span, are to be checked to determine the need for reinforcement due to rearrangement of the stress–strain state parameters in their structural elements as a result of a change in the static behavior pattern. If, according to the results of the survey, the need for strengthening is identified, then work is to be carried out to implement it; however, preliminary calculation established that these spans usually will not require to be strengthened. And finally, at the third step, instead of the destroyed part of the five-span structure, a new single-span structure is installed (Fig. 12b; gray—existing undamaged spans, black—reconstructed spans).

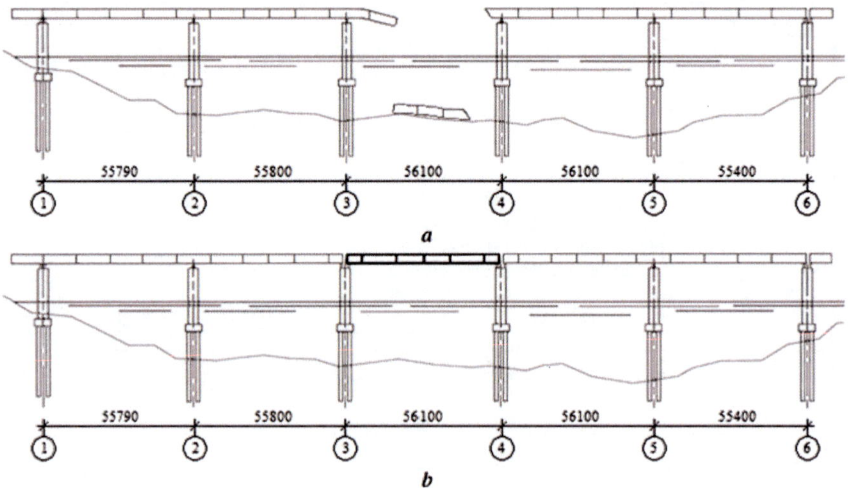

Fig. 12 Schematic structural diagrams of the most likely destruction and reconstruction of the road part of the Darnytsky rail-road bridge: **a** superstructure after destruction of span 3–4; **b** superstructure after reconstruction

Two design solutions of a new single-span structure are considered, the first of which is a steel–concrete superstructure and the second one is a steel girder superstructure with an orthotropic deck (Fig. 13). Comparison of the proposed solutions showed that the second of them is more rational, which is distinguished not only by its high manufacturability and process-ability for installation, but also by the accelerated rates of work with minimal labor costs. In this case, it is very important that all structural members of the steel girder superstructure with the orthotropic deck can be pre-fabricated in the factory. The only drawback of the second variant of the new superstructure is its somewhat higher cost compared to the first variant. However, according to the authors, all the existing advantages of the second design solution practically eliminate this drawback, especially if we take into account the usual strict requirements for terms of eliminating the consequences of terrorist attacks.

4 Conclusions

Probable variants of terrorist attacks on the Paton Bridge across the Dnieper River are considered, the methods for eliminating their consequences are developed and four most likely schemes for destruction of bridge structures are established. Within the framework of these methods, three variants are proposed for replacing damaged sections of spans. Taking into account considerations regarding the accelerated rates of work with minimal labor costs, high manufacturability and installation process-ability of structures, preference is given to metal structures. A lightweight steel superstructure with a carriageway in form of steel orthotropic deck and with two

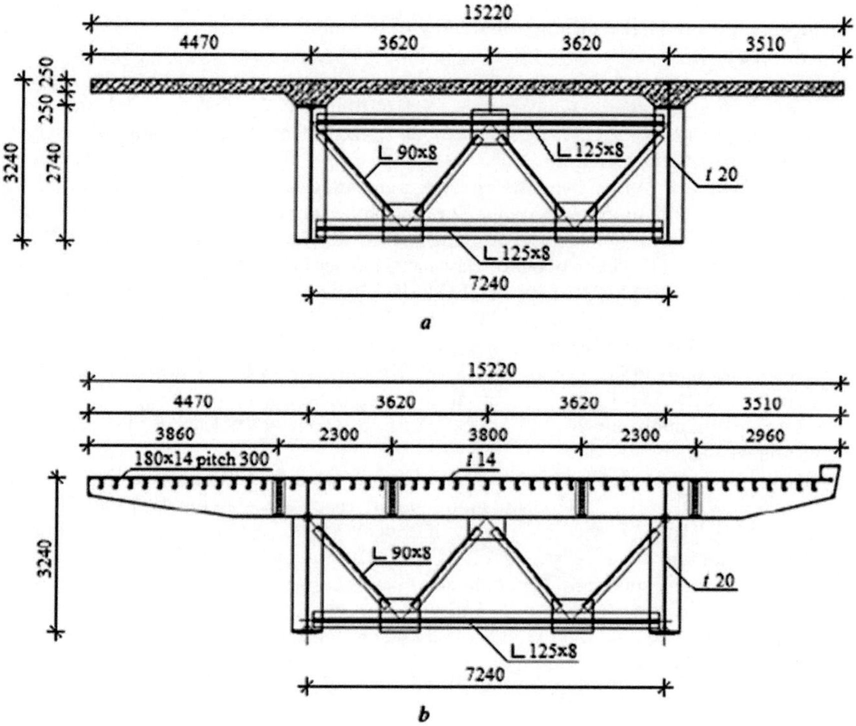

Fig. 13 Design solutions of new single-span structure: **a** steel–concrete span structure; **b** steel girder span structure with orthotropic deck

main girders located at a distance of 11.9 m from each other is recommended for use. It is proposed to rebuild damaged piers using steel lattice structures made of extensional units.

Probable variants of terrorist attacks on the Darnytsky Rail-Road Bridge crossing over the Dnieper River are considered, the methods for eliminating their consequences are developed and the most likely scheme for destruction of bridge crossing structures is established. Within the framework of these methods, two design solutions are proposed for replacing damaged section of the superstructure. It is recommended to use a steel girder superstructure with an orthotropic deck, which differs not only by high manufacturability and process-ability for installation, but also by accelerated terms of work with minimal labor costs.

References

1. Шимановський О.В. Особливості убезпечення будівель та інженерних споруд при терористичних нападах // Промислове будівництво та інженерні споруди.

2018. N 4. С. 2–11 (Peculiarities of security of buildings and engineering works in view of terrorist attacks)
2. Шимановський О.В. Методи та практичні заходи з ослаблення наростаючого руйнування сталевих конструкцій // Промислове будівництво та інженерні споруди. 2019. N 1. С. 2–14 (Methods and practical measures to mitigate cumulative destruction of steel structures)
3. Шимановський О.В. Сучасний технічний стан позакласних металевих мостів м. Києва // Збірник наукових праць Українського інституту сталевих конструкцій імені В.М.Шимановського. – Київ: Видавництво "Сталь". 2018. Вип. 21–22. С. 4–37 (Current technical condition of out-of-class metal bridges in Kyiv)
4. Шимановський О.В., Котлубей Д.О., Шалінський В.В. Міст ім. Є.О.Патона – нинішній стан та перспективи // Промислове будівництво та інженерні споруди. 2018. N 1. С. 2–9 (PatonBridge—currentstateandprospects)
5. Шимановський О.В., Котлубей Д.О., Шалінський В.В. Аварійна ситуація на мосту ім. Є.О.Патона та заходи щодо її вирішення // Промислове будівництво та інженерні споруди. 2018. N 4. С. 30–33 (Exigent condition of Paton Bridge and remedial measures)
6. Шимановський О.В., Котлубей Д.О., Шалінський В.В. Щодо технічного стану моста ім. Є.О.Патона // Промислове будівництво та інженерні споруди. 2019. N1. С. 20–25 (Regarding serviceability of Paton Bridge)
7. Шимановський О.В., Котлубей Д.О., Шалінський В.В. Технічний стан автодорожньої частини Дарницького залізнично-автомобільного мостового переходу через р. Дніпро у м. Києві // Промислове будівництво та інженерні споруди. 2019. N 2. С. 8–13 (Serviceability status of road part of Darnytsky rail-road bridge crossing over Dnieper River in Kiev)

Influence of Major Beam Damage on the Redistribution of Internal Forces in Concrete Road Viaducts

K. Jurasz-Drozdowska and W. Radomski

Abstract This article describes the problem of redistribution of internal forces that occurred as a result of damage to the bridge structure girders caused by the impact of an oversized vehicle. In the Sofistik computer program, a three-span five-girder and seven-girder viaduct was modeled. The structure of damage to the bottom of the structure as a result of vehicle impact in the middle of the span of the second span was determined in the object. Static and strength analyzes of the viaduct structure were carried out, comparing the results obtained for the original condition and the damaged condition.

Keywords Bridge · Damage to girders · Redistribution of internal forces

1 Introduction

The term redistribution comes from the two Latin words 're'—backwards, opposite, opposite and 'distributio'—distribution, arrangement. This term can be represented as a repetition, re-division, decomposition of something. Over the years, bridge structures need modernization and repair. They should be adapted to new communication conditions, and thus increase their load capacity or extend the usable width of the road. Any interference resulting from structural and functional modernization is always associated with a greater or smaller redistribution of internal forces in the structure.

The phenomenon of internal forces redistribution can have many causes, from the very beginning of the erection of the object, in the assembly phases, to renovations and modernizations. Redistribution may also result from improper use of the structure. Along with the increase in communication in recent years, oversize and oversized

K. Jurasz-Drozdowska (✉)
Faculty of Civil Engineering and Architecture, Opole University of Technology, Opole, Poland
e-mail: k.jurasz-drozdowska@po.edu.pl

W. Radomski
Faculty of Civil Engineering, Warsaw University of Technology, Warsaw, Poland

Z. Zembaty et al. (eds.), *Environmental Challenges in Civil Engineering*, Lecture Notes in Civil Engineering 122,
https://doi.org/10.1007/978-3-030-63879-5_3

vehicle traffic has increased on the roads. By definition, the route of such vehicles should be planned in detail, adapted to the road gauge and agreed with the relevant authorities and institutions. Often, however, the carelessness and bravado of drivers and carriers can lead to a threat to traffic safety as well as the safety of the engineering facilities themselves, which are on the route of an oversized vehicle. The situation in which the object is damaged by a car impact is one of many that during the operation of the object may cause a sudden redistribution of internal forces.

Bridge failures arouse quite wide offers, but there are no compact and comprehensive publishing positions about this list. On a global scale, there are several extensive items describing bridge failures such as D.W analysis. Smith from 1977 [1] concerns 143 bridge disasters and a newer item from 2010 [2]. These deficiencies are available from some confidentiality investigation documents regarding the actual causes of the merger or disaster, and will be subject to their consequences, requesting transactions and financial advice. They are often "blanked" and it is difficult to obtain reliable information about them. However, it is worth reviewing references to the last two Conferences in Construction Failures to get to the conclusion that this problem is not at all marginal [3–6]. In the list of cases, these are only descriptions of the situation together with a description of the extent of connections and corrective actions [4]. Less often, analysis of span analysis attempts with reports confirmed by information [3, 6].

However, several descriptions of in-depth analysis appeared in the world literature, including modeling of the phenomenon of collision with a detailed model of a striking vehicle and a non-linear model of the bay [7, 8]. This simulation makes it possible to assess the degree of danger in the event of a vehicle colliding with a bridge span, predicting the effects of such an event, and seeking and designing appropriate safeguards.

This article deals with the problem of redistribution of internal forces that occurred as a result of damage to the bridge structure girders caused by the impact of an oversized vehicle. Using the Sofistik computer program, a three-span viaduct in a five-girder and seven-girder variant was modeled. The structure of damage to the bottom of the structure as a result of vehicle impact in the middle of the span of the second span was determined in the object. Static and strength analyzes of the viaduct structure were carried out, comparing the results obtained for the original condition and the damaged condition.

2 Redistribution of Internal Forces in the Exploitation Stage

During their operation, bridge structures are exposed to various phenomena that may affect the redistribution of internal forces. Along with the increase in communication in recent years, oversize and oversized vehicle traffic has increased on the roads. By definition, the route of such vehicles should be planned in detail, adapted to the road gauge and agreed with the relevant authorities and institutions. Often, however, the carelessness and bravado of drivers and carriers can lead to a threat to traffic safety as well as the safety of the engineering facilities themselves, which are on the route of an oversized vehicle. This situation is one of many situations that during the operation of the facility may cause a sudden redistribution of internal forces. Damage to the structure of the bridge structure may be associated with economic effects in the form of costs of repair, possible demolition or reconstruction, but also with social effects [9], the scale of which can be discussed depending on the communication significance of the damaged structure.

An example of such an event is described in article [10], which presents damage to the structure of the viaduct located in the town of Borkowo above the Tri-City beltway, which is one of the most important communication routes in the Pomeranian Voivodeship.

The supporting structure of the viaduct span was damaged as a result of an oversized vehicle of an oversized vehicle—an excavator transported on a tow truck. Three of the four main girders in the central part of the second span were significantly damaged (Figs. 2 and 3a). As a result of the impact with the span No. 2, practically three of the four girders disengaged from the work in the central part, changing the static (work) scheme of the entire facility (Fig. 3b). Damaged girders adopted the diagram of suspended brackets and the weight of this part of the structure was redistributed to the new static system of the object. Both girders damaged in their other parts (support sections) as well as girders in neighboring spans suffered overloads. Immediately—until static and strength analyzes were carried out—the object was decommissioned by completely preventing traffic on it. In addition, the structure of the damaged span was supported on temporary supports.

In connection with the emergency condition of the facility, static and strength calculations of the viaduct were carried out [11]. The following variants of the object's calculations were considered:

- undamaged condition—to determine the level of internal forces and stresses for which the object was designed,
- damaged condition—the damages of the structure of the span No. 2 resulting from the impact were reproduced—internal forces giving the image of strain of the damaged structure were determined,
- condition after dismantling of bay 2—the level of internal forces in the remaining part of the object was analyzed in a changed static scheme compared to the original,
- states reflecting the individual stages of disassembly of bay 2 in order to check the level of structural effort, and in particular—pillar transoms.

The static and strength analyzes of the viaduct structure showed, in accordance with previous assumptions, that as a result of vehicle impact the static system of the object changed significantly [11]. Virtually completely destroyed three girders in the middle zone of bay 2 adopted a diagram of brackets and a redistribution of weight took place throughout the entire facility (Fig. 3a, b) In the changed static system under the influence of constant loads, taking into account possible temperature changes and snow retention, the system was overloaded load-bearing viaduct both in the support zones of damaged girders as well as girders in neighboring spans. Internal forces in the transom beams also increased, which, however, did not overload these elements.

Not only the impacts on the bottom of the structure can cause the redistribution of internal forces. All other collisions of means of communication with the bridge object can contribute to the impact on its work, we are talking here about bridges over navigable rivers, where there is a risk of the ship being hit by intermediate supports, as well as the boughs etc. carried by water. Attention should also be paid to the movement on the site, because in the case of structures with downhill driving (e.g. arch bridges, truss) there is a risk of collision of cars with the main girder. That is why it is necessary to properly design equipment components, including energy-intensive barriers, aimed at preventing this type of failure, because when this happens, it is not enough to remove damaged vehicles, but also to properly examine the structure to confirm its ability to continue operation.

3 Computational Models

The article deals with the problem of the impact of sudden damage to the bridge on its condition and the possibility of further operation. The fictitious three-span bridge with five main girders has been analyzed (Fig. 4). The structure of the underside of the structure as a result of the impact of the oversized vehicle in the middle of the second span has been modeled. Damage in the longitudinal section is distributed as follows: extreme girder No. 1 damaged over a width of 5 m, successively girder No. 2 over a width of 3.5 m and girder No. 3 over a width of 2 m (Fig. 1). Figure 5 presents the course and dimensions of damages of individual main beams in transverse view. The course of the damage line is similar to the one in Fig. 3b. The object was modeled using the finite element method (FEM) using the SOFiSTiK program as a beam-shell model (Fig. 1). The bridge was introduced to the program by means of a problem-oriented language of CADINP preprocessors in alphanumeric format, introduced in the TEDDY text editor—this is the basic way to declare data in the system. The structure was analyzed and calculated for a class B load [12].

The analysis consisted in comparing the results obtained after calculating two bridge construction systems:

(a) three-span, 5-girder bridge, (output system)
(b) three-span, 5-girder bridge, system with modeled damage, formed in the middle of the span as a result of impact of an oversized vehicle (damaged system). The above cases were additionally considered in two variants. The first assumed a continuous structure, the second with two joints in the middle span. So in summary, the following cases were considered for a 5-girder structure:

- bridge without damage (marking—m1),
- bridge with damage (marking—m1u),
- bridge without damage, with joints (marking—m2p),
- bridge with damage, with joints (designation—m2pu).

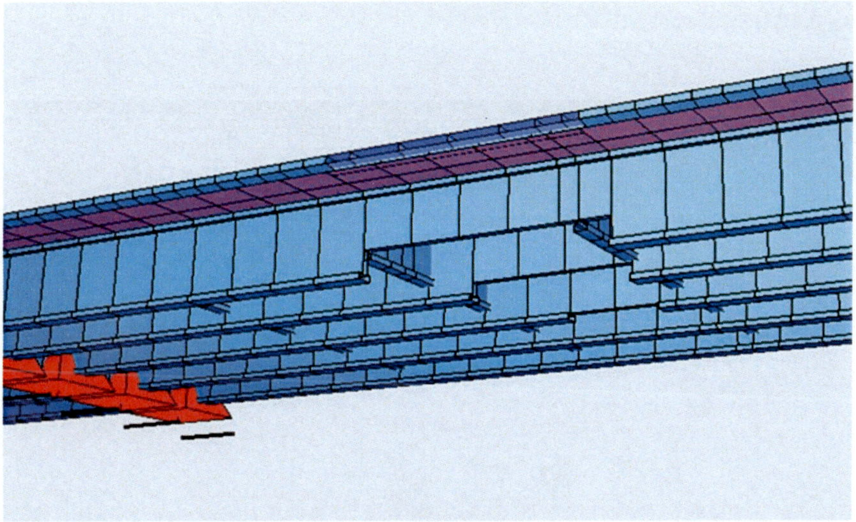

Fig. 1 A bay with a modeled damage to the main girders

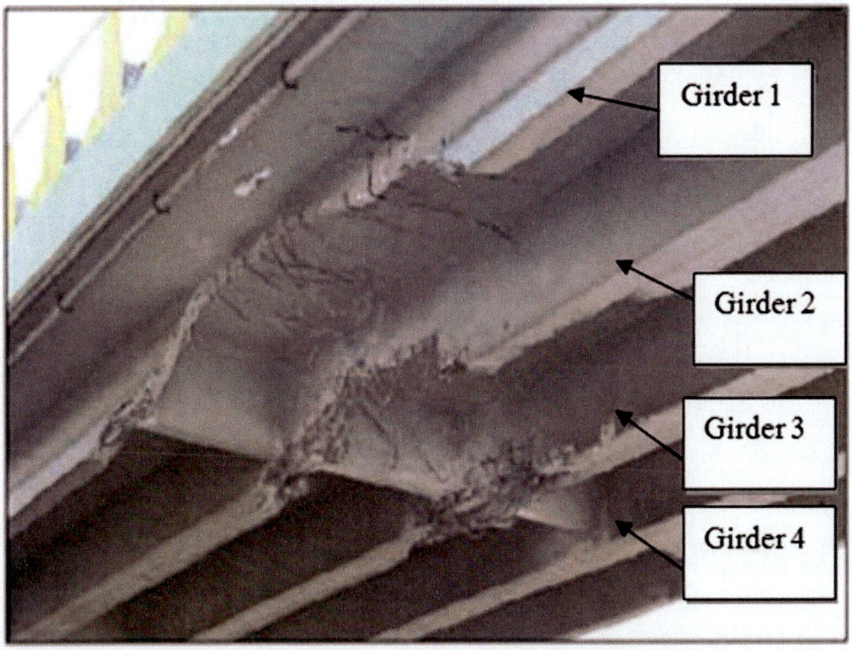

Fig. 2 The scale of damage to the main girders of the damaged bridge structure [10]

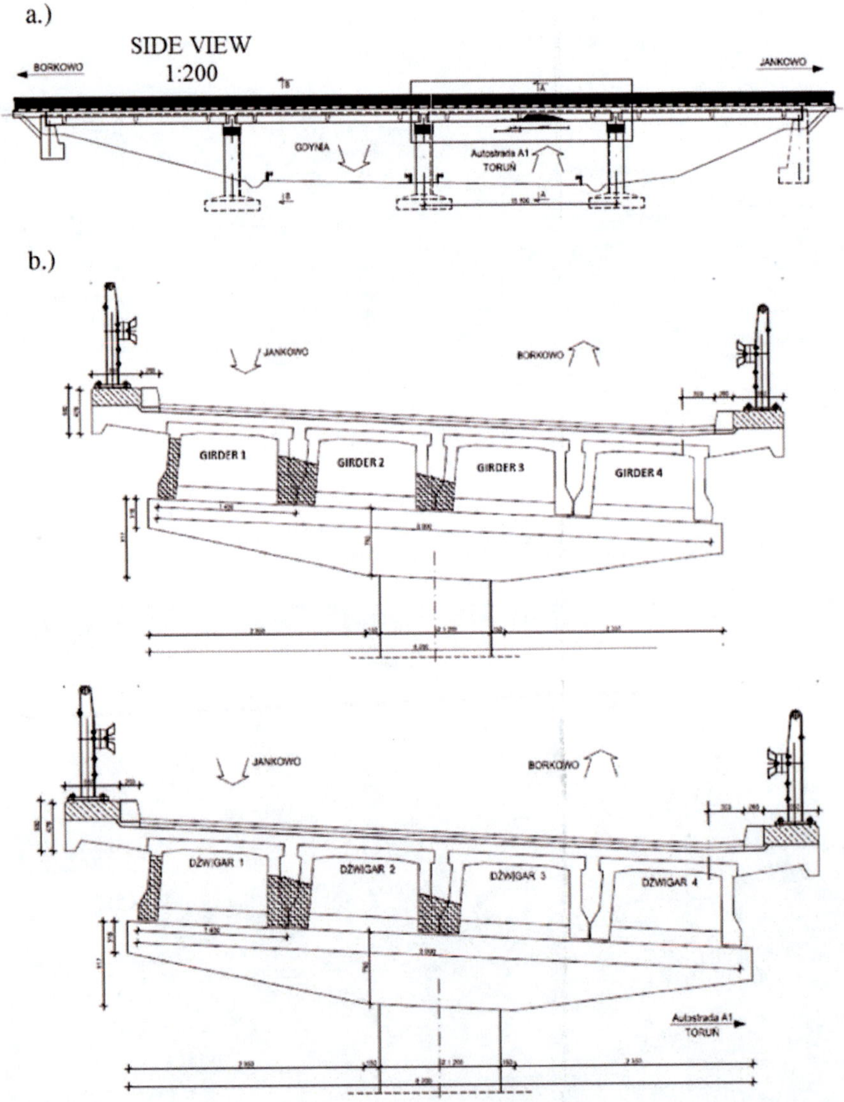

Fig. 3 Scale of damage to the main girders; **a** longitudinal section of the bridge, **b** cross-section of the bridge [11]

Below the graphs (Fig. 6) show the values of bending moments obtained by means of numerical calculations for the first girder No. 1. Because for each of the beams such charts were obtained, with the values of moments, the authors decided to include only this for the most damaged girder (Fig. 6). For the remaining girders, the following relationships are presented in the following sections to present the results obtained in a more vivid and transparent way. The results of the bending moments

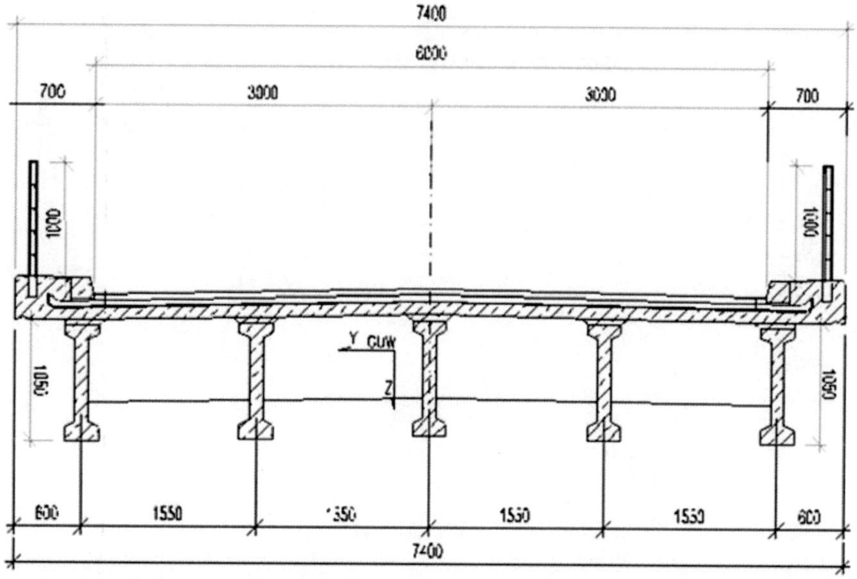

Fig. 4 Cross section of the bridge span

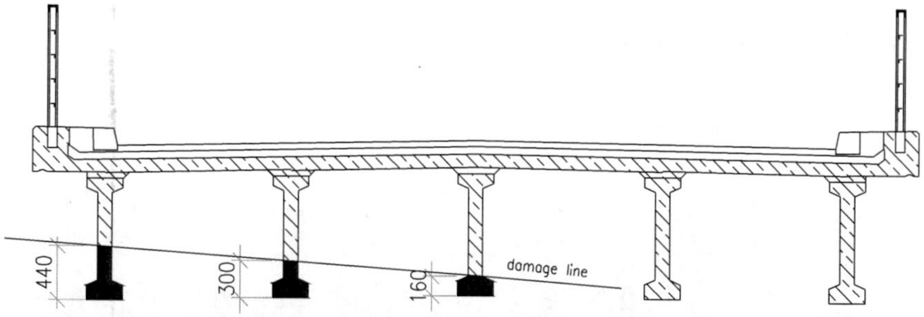

Fig. 5 A drawing showing a diagram of a bridge damage in cross section

obtained from the "a" model were determined as the initial ones to which the bending moments obtained from the "b" model (damaged system) were compared. Thus, the results from the "a" model received 100%, and the results from the "b" model were calculated against these 100% to show the change in moments after the damage was introduced.

3.1 Girder Number 1

The results obtained for the 1/5 girder (Fig. 7) clearly show a significant decrease in the span moment in the middle span, and smaller decreases in the extreme spans.

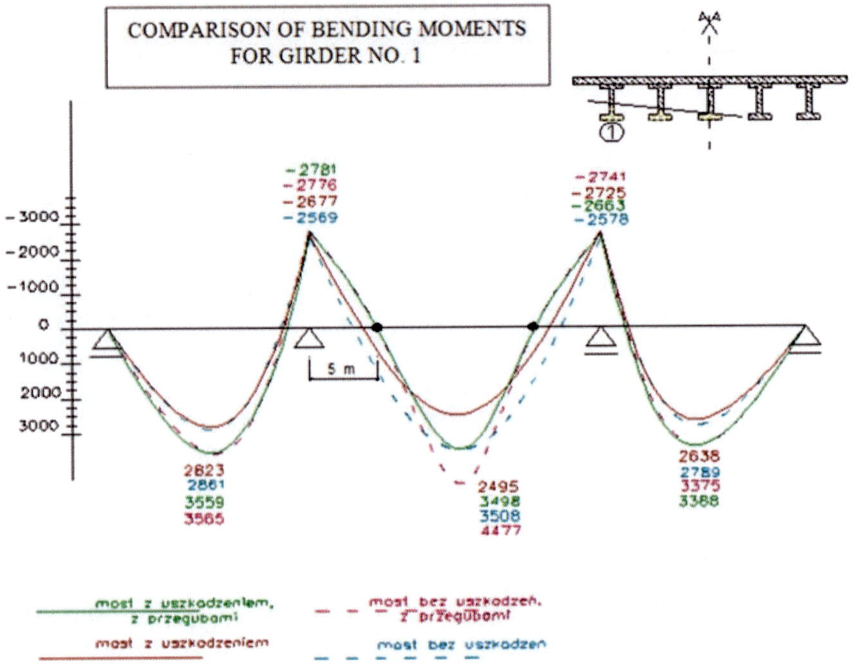

Fig. 6 List of moments and cutting forces for the most damaged girder No. 1

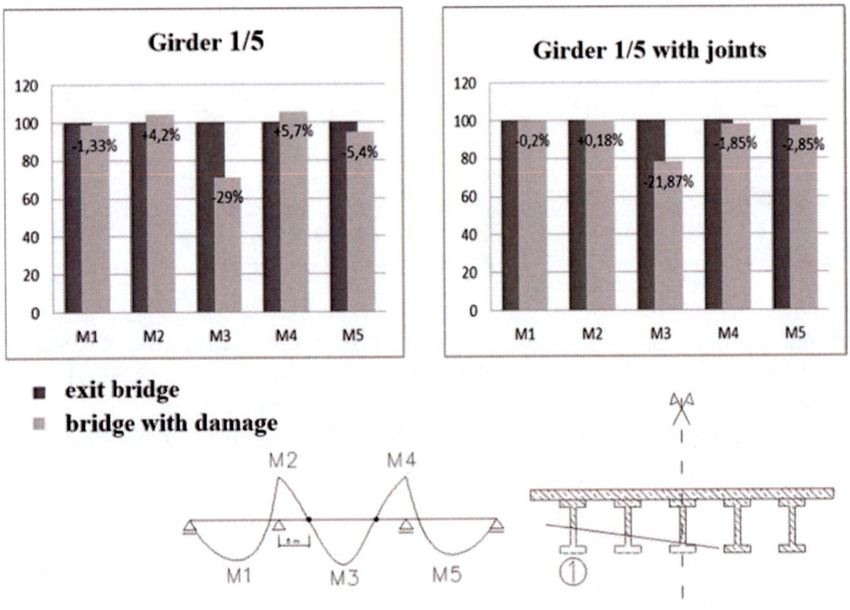

Fig. 7 Charts indicating the percentage differences in the results obtained for girder No. 1

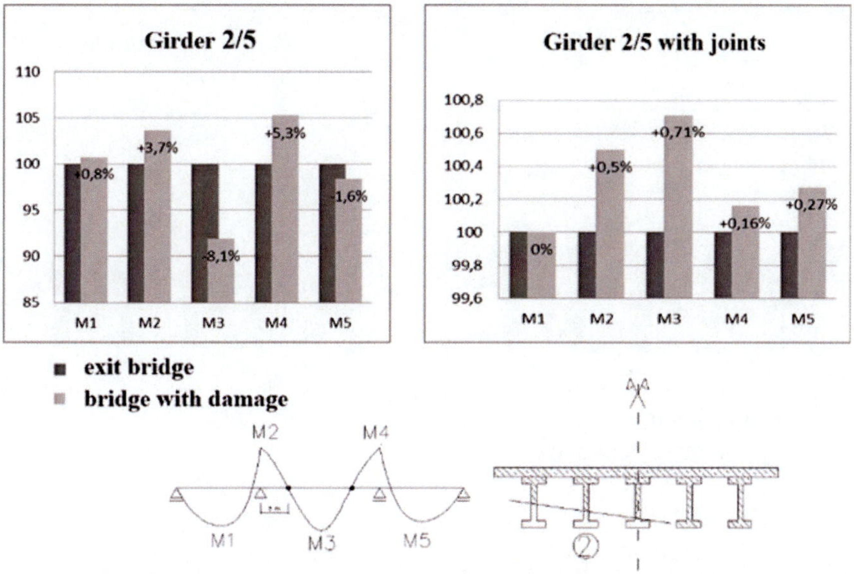

Fig. 8 Charts indicating the percentage differences in the results obtained for girder No. 2

Comparing the two graphs, it can be seen that in the articulated system the reduction of span moments is smaller, and the support moments change not as much as in a continuous system.

3.2 Girder Number 2

The results obtained for girder number 2, presented in Fig. 8, show a clear increase in the value of support moments for a bridge without hinges and a decrease in span moments. In the case of constructions with hinges in the middle span, the moment values increased slightly.

3.3 Girder Number 3

Analyzing the results obtained for girder number 3 (Fig. 9), there is a significant increase in the span moment in the middle span, both in constructions with and without joints. The support moments M2 and M4 also increased (smaller increase in the articulated bridge). Spanning moments M1 and M5 slightly increased in the object with joints, and their value dropped in the second structure under consideration.

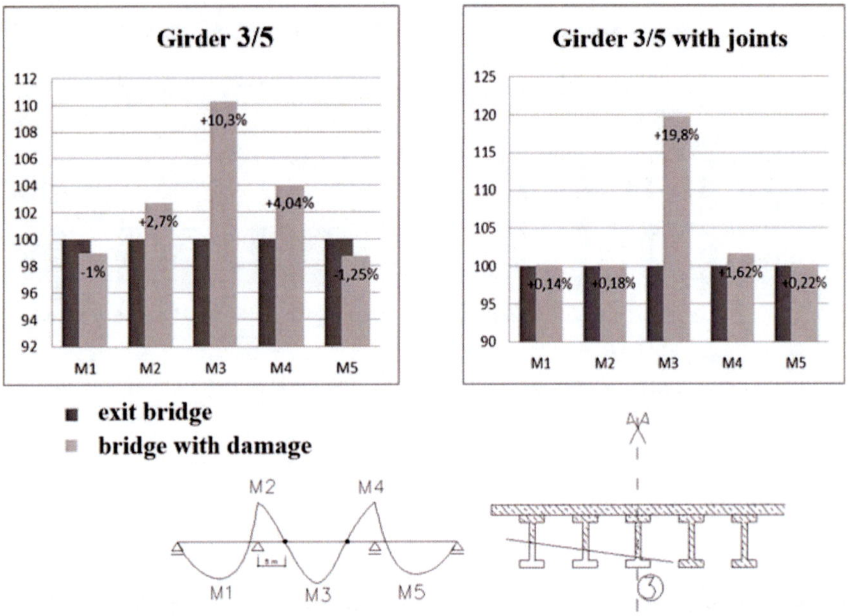

Fig. 9 Charts indicating the percentage differences in the results obtained for girder No. 3

3.4 *Girder Number 4*

The results obtained for girder number 4 show a similar situation as for girder 3/5. Significant increase and decrease of the moment value occurs in the same places (Fig. 10).

3.5 *Girder Number 5*

Analyzing the results obtained for the second extreme girder number 5 we notice (Fig. 11) that the arrangement of new moments looks similar for the two constructions considered: a bridge with joints and without joints. Practically in every case there was a decrease in the moment value, with the largest occurring in the middle span M3, a smaller decrease by approx. 2% compared to the initial value can be seen at the support moments M2 and M4. A slight change of approx. 0.5% can be seen in the extreme spans M1 and M5.

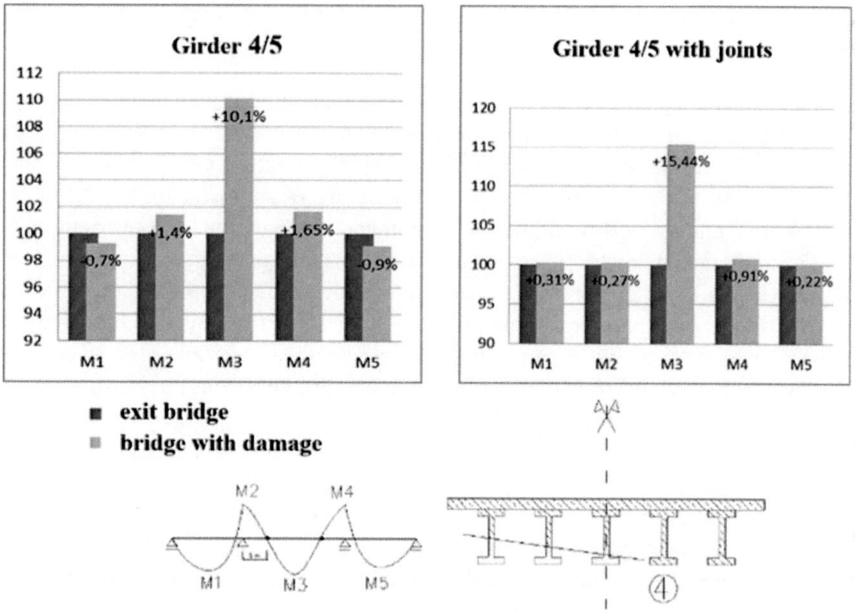

Fig. 10 Charts indicating the percentage differences in the results obtained for girder No. 4

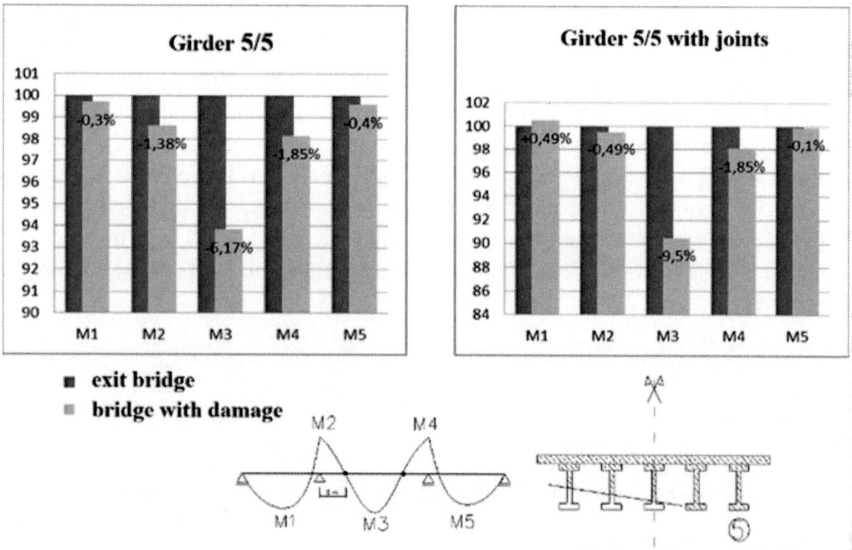

Fig. 11 Charts indicating the percentage differences in the results obtained for girder No. 5

4 Summary

Impact damage to bridge structures is becoming more common in practice. In engineering practice, it introduces safeguards in the form of extreme gates limiting the object, traffic lights, as well as warning signs. Unfortunately, even such actions are not able to stop drivers from trying to "fit" under the object, which is the result of a lack of imagination of drivers and total ignorance of regulations and restrictions.

The numerical analyzes presented in the article, made using the Sofistik program, show how the work of the bridge structure changes in the event of damage to the girders. In the calculation example, damage to three main beams was modeled, the scale of which is not as large as in the case described in article [10] (where the extreme girder was destroyed along its entire height). However, this was enough to show what disorder occurs at bending moments. The most overloaded was girder No. 3, located in the middle of the span (damage to the bottom shelf) and girder number 4 that was not damaged. These girders were loaded with higher values of bending moments in relation to the initial model (without damage). It is worth noting that the girders in the articulated system are more overloaded, and there are many such bridge structures in Poland. As a rule, these are constructions of several decades, often built using these joints in their arrangement.

The next stage of the work will be the analysis of the impact of damage on the operation of the bridge construction depending on the location of this damage. The article assumes that the failure occurred in the middle of the second span, but additional analysis would be diversified by "shifting" this damage to different places of the structure, e.g. at the intermediate support or in the extreme pass. The topic is very broad and gives further scope for further analysis. However, in any case where there is structural damage resulting from the impact of an oversized vehicle, immediate safety measures must be taken. The effects of this type of bridge failure are both economically and socially significant. It should be assumed that the building is in a state of emergency and is not suitable for temporary use, or in extreme cases a decision will be made to close the building and the need to temporarily support the damaged span.

References

1. Scheer J, Wilharm L (2010) Failed bridges: case studies, causes and consequences
2. Smith DW Why do bridges fail, 1977-11, Civil Engineering, American Society of Civil Engineers. ISSN 0885-7024
3. Bętkowski P, Pradelok S (2007) Uszkodzenie i tymczasowa naprawa typowego kratowego mostu kolejowego. In: XXIII Konferencja Naukowo-Techniczna Awarie Budowlane, Szczecin-Międzyzdroje, 23–26 maja 2007
4. Budka E, Lewandowski G, Kosuch M, Lorenc W, Rabiega J (2009) Przykłady uszkodzeń konstrukcji mostowych spowodowanych uderzeniami pojazdów. In: XXIV Konferencja Naukowo-Techniczna Awarie Budowlane, Szczecin-Międzyzdroje, 26–29 maja 2009

5. Radomski W (2007) Dynamiczne przyczyny awarii mostów. In: XXIII Konferencja Naukowo-Techniczna Awarie Budowlane, Szczecin-Międzyzdroje, 23–26 maja 2007
6. Żółtowski K, Szafrański M (2009) Uszkodzenie wiaduktu kolejowego nad drogą krajową nr 7. Analiza nośności i sposób naprawy. In: XXIV Konferencja Naukowo-Techniczna Awarie Budowlane, Szczecin-Międzyzdroje, 26–29 maja 2009
7. Xin-zheng L, Yan-sheng Z, Jian-jing J, Ai-zhu R, Jing N (2007) Nonlinear finite element simulation for the impact between over-high truck and bridge-superstructure. In: Proceedings of 7th international conference on shock and impact loads on structures, Beijing, pp 387–394
8. Yang M, Qiao P, McLean D, Khaleghi B (2010) Effects of overheight truck impacts on intermediate diaphragms in prestressed concrete bridge girders. PCI J 55:58–79
9. Radomski W (2003) Społeczne skutki awarii mostów—przykład warszawski. In: XXI Konferencja Naukowo-Techniczna "Awarie Budowlane", Szczecin—Międzyzdroje, 20–23 maja 2003
10. Chróścielewski J, Banaś A, Malinowski M (2013) Damage of road viaduct above the bypass of Trójmiasto. Inżynieria i Budownictwo 69(7–8)
11. Chróścielewski J, Malinowski M, Banaś A, Sitarski A (2011) Ekspertyza techniczna dotycząca uszkodzenia wiaduktu drogowego nr 17 nad obwodnicą Trójmiasta w km 349 + 230 w m. WILiŚ. KMBiM, Gdańsk, Borkowo, Politechnika Gdańska
12. PN-85/S-10030 Obiekty mostowe. Obciążenia

Pull-Out Tests and Numerical Simulations of Bonded Threaded Anchors in Concrete Blocks with Thermal Insulation Layer

D. Tomaszewicz, A. Jablonska-Krysiewicz, and M. Gryniewicz

Abstract The article presents the results of pull-out tests and computer simulations of three-layer concrete block elements with an internal insulation layer made of foamed polystyrene. The concrete block elements are fragments of an external layer of a wall of traditional large-panel buildings. The purpose of the tests was to obtain the ultimate destructive force of a connection. It is not desirable to check the carrying capacity of such anchorages in a residential building. E.g., pull-out tests are the destructive ones and can cause permanent damage to a wall. The durability of a connection surface between the textured layer and other ones strictly depends on the durability of near anchoring. The concrete block elements were designed using C 12/15 class concrete in accordance with Eurocode 2. The connection of a threaded steel anchor with concrete was made using "R-KER vinyl ester bonded anchor" resin with an adhesion stress value of 7 N/mm2. The steel anchor is a threaded rod with a diameter of Ø 12 mm made of steel grade 5.8. The computer simulations are a conceptual proposal for modeling layered block elements in systems and connections not used so far. A primary aim of them was a validation of the experimental tests and a first attempt to the theoretical assessment of the single adhesive anchor strength. In the numerical modeling, it was necessary to accurately model contact in joints, like resin with concrete or resin with steel anchor. The assumed thickness of the connection area is 2 mm in the computer simulations (theoretical space between a concrete and an anchor outer surface in a connection). The threaded part of rods is taken into account in FEM calculations.

Keywords Bonded anchors · Numerical simulation · Concrete block elements

D. Tomaszewicz (✉)
Higher School of Agribusiness in Lomza, Studencka 19, 18-402 Lomza, Poland
e-mail: Dariusz.Tomaszewicz@poczta.wsa.edu.pl

A. Jablonska-Krysiewicz · M. Gryniewicz
Faculty of Civil and Environmental Sciences, Bialystok University of Technology, Wiejska 45E, 15-351 Bialystok, Poland

Z. Zembaty et al. (eds.), *Environmental Challenges in Civil Engineering*, Lecture Notes in Civil Engineering 122,
https://doi.org/10.1007/978-3-030-63879-5_4

1 Introduction

The tested concrete block elements represent fragments of an external layer of a wall of traditional large-panel residential buildings where it is not desirable to check the carrying capacity by pull-out tests. Studies on multilayered walls are important, i.e. where the outer layer is not sufficiently load—bearing to transfer the load from the anchor or the outer layer has to be supported by the internal one. The tests described here were undertaken to preliminary inspect the behavior of such structures. The results were generally verified using simplified computational simulations, which also can be treated as a proposition for the future low-cost theoretical assessment of the load capacity of multilayered walls connections. Numerical modeling of bonded anchorages in concrete is done by many authors [1–9], as well as in wood [10]. Some cases relate to testing the reinforcement load capacity [11, 12]. All numerical analyzes are preceded by more or less complex experimental studies [13–17]. The authors of this article have included in the numerical calculations the special feature function defined for the threaded connection of a screw anchor. Due to this parameter, it was not necessary to model the actual geometry of the anchor.

2 The Description of the Experiment

A stepwise increase in the pull-out force was applied during the loading. It was planned before conducting experimental tests, using the dedicated software operating the HYSDOZOK hydraulic load system (Fig. 1) [18, 19].

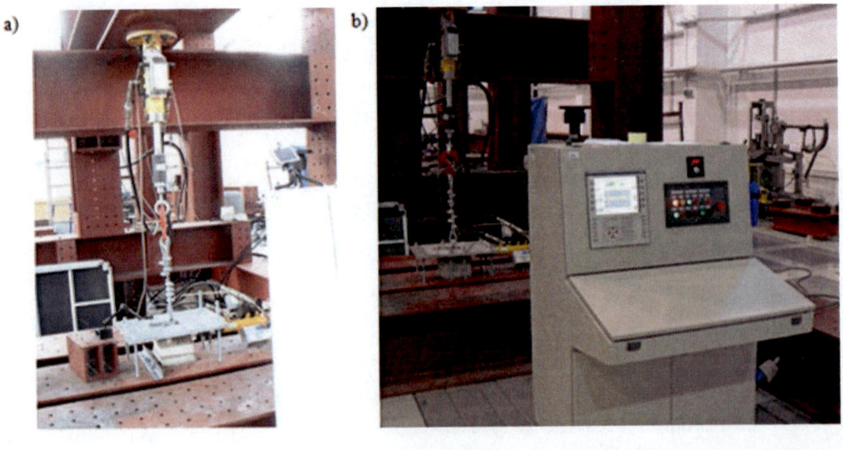

Fig. 1 Set up of the experimental tests just before start: **a** a three-layer solid element with an adhesive anchor; **b** the HYZDOZOK operating system and a complete solid concrete element with a concrete anchor

To conduct these tests, a test stand was made, which consisted of 1 cm thick resistance slab with 80 × 40 cm cross-sectional dimensions with a 10 × 10 cm cut-out the hole in the center. This allowed free pulling of the bolt with any shape of a broken concrete cone. Other parts of the stand were 4 or 6 pieces of an M20 screws fixing the resistance slab with to bottom beam. The rope with a diameter of Ø13 mm was ended on one side with a hook with a load capacity of 3 tons. On the other site it was ended with an M12 eye, which was predicted to screw the anchor. In order to attach the hook to the hydraulic cylinder, a double-sided thread was used: M22 mounted in the hydraulic cylinder and M24 ended with a threaded eye. Additionally, the rollers between the resistance slab plate and the upper surface of a specimen has been added to eliminate the secondary frictional forces caused by horizontal (shear) force. In order to block a sample movement in a bottom area, the steel angle was screwed to the stand beam, with a roller placed between it and the sample.

In the first stage of tests, anchors fixed at an angle of 90° in relation to the concrete surface were tested (Fig. 1). They were both bonded anchors (Fig. 1a) and fixed in complete solid concrete elements (Fig. 1b).

Additional tool plates with an opening were attached to the anchors and screwed to help in displacements measurements. The first stage of the research proceeded as follows: for the previously prepared auxiliary equipment, a vertical hydraulic cylinder was installed in HYSDOZOK to obtain a pulling force, programmed in an experimental study to achieve the final limit of 80 kN strength. The results reading has been done in steps of 5 kN with a steady increase in force of 0.5 kN/s. The shear force in the case of three-layer samples, has been established using a unit load of 1.4 kN corresponding to a 1 m^2 of the surface of the texture layer. For testing oblique anchors in the 2nd stage of testing, the test stand (Fig. 2) was additionally equipped with racks, made at the expected angles on which subsequent samples were placed. Diagrams of test stands are shown in Figs. 3 and 4.

Samples with dimensions: 20 cm × 20 cm × 17 cm were tested, in the case of three-layer samples, the cross section thickness of 17 cm is: 6 cm construction layer, 6 cm insulation layer, 5 cm textured layer and 30 cm × 20 cm × 6 cm for anchoring at 30°. The conditions for supporting solid elements are shown in Figs. 5 and 6. The material constants contained in Table 1 were adopted for testing and calculations.

3 Numerical Modeling of the Block Elements

Numerical modeling of block elements was performed in ANSYS Workbench. One of the most complex processes was the proper design of contact zones in each model. Modeling of block elements using the Finite Element Method was carried out with all parameters and material constants (Table 1) taken into account during the experimental research. The assumption was to build the simplified, practical model suitable for the experimental validation and proposition for a future general assessment of the load capacity of multilayered walls connections. The concrete, steel and resin materials were described using bilinear relationships. Nonlinearity caused by cracking

a) b)

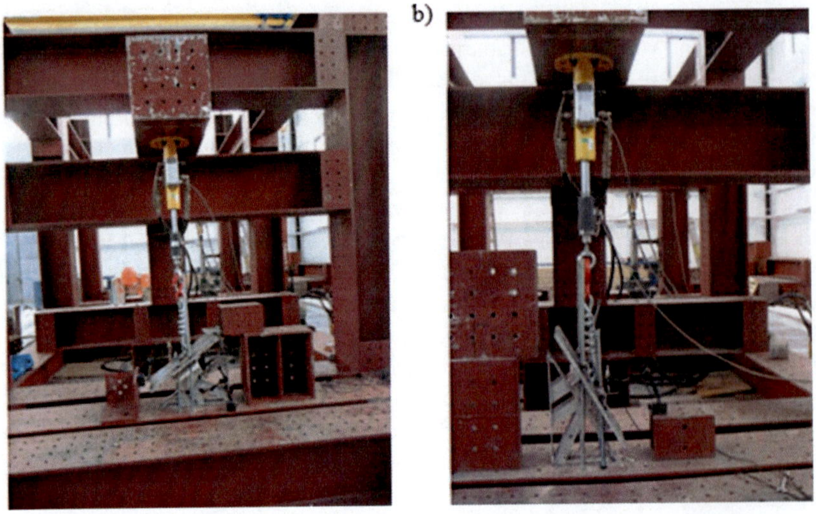

Fig. 2 A view of experimental tests just before starting: **a** testing a three-layer solid element with an anchor bonded at a 45° angle; **b** testing of a single-layer solid element with an anchor bonded at an angle of 30°

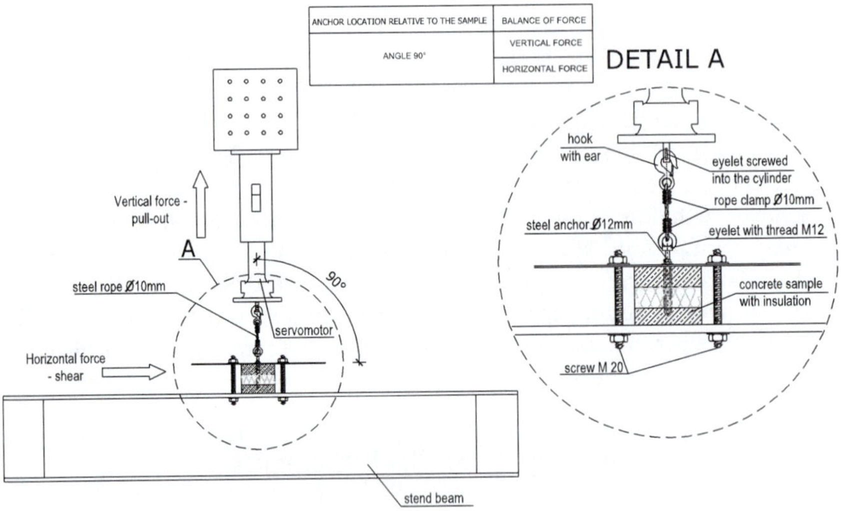

Fig. 3 Scheme and research procedures in the planned research [20]

and crushing of concrete elements were not analyzed here. This is the subject for future detailed studies after the overview analyses. The model (Figs. 7 and 8) of the block element consists of the three parts: a bonded anchor solid, concrete blocks and an insulation layer. The primary contact zones are: the bonded anchor—the

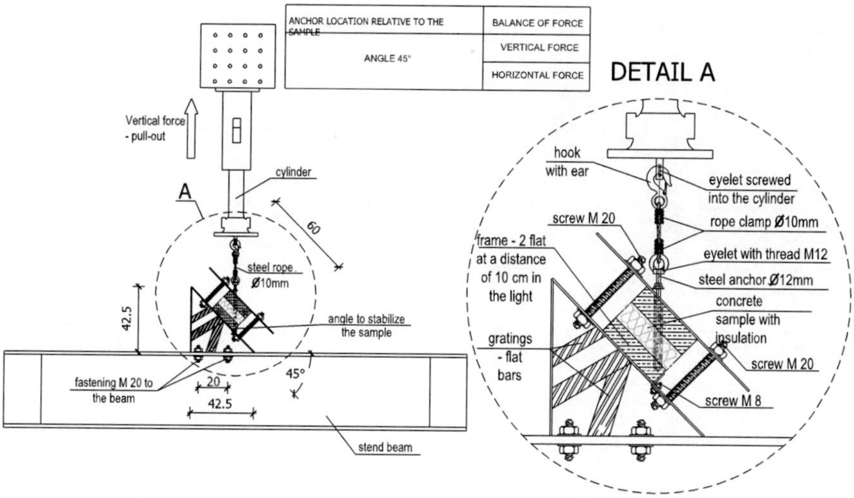

Fig. 4 Scheme of the test bench for oblique anchors at an angle of 45° [20]

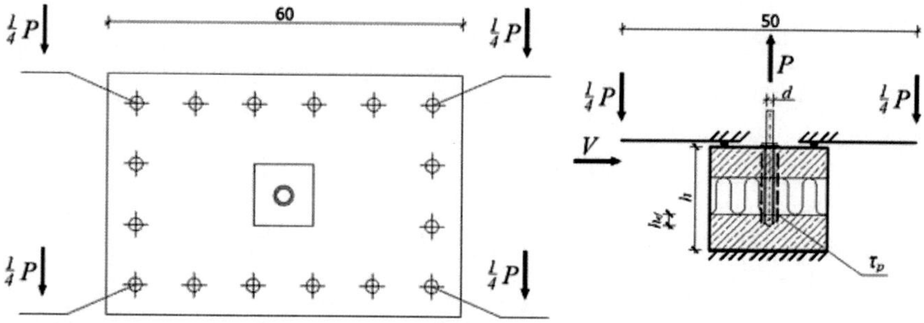

Fig. 5 Scheme of three-layer sample load with anchor bonded perpendicularly

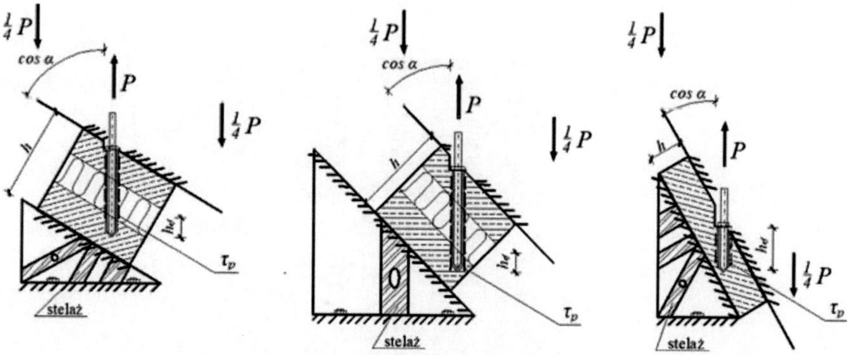

Fig. 6 Sample load diagram with single diagonal anchors at 60°, 45° and 30° respectively

Table 1 Summary of results of experimental and numerical calculations for the second model (bonded anchors at an angle of 30°)

Material parameters	Material name			
	Concrete C 12/15 Steel grade 5.8	Insulation layer	Epoxy resin	(contact layer)
Density (kg m^{-3})	2300	7850	950	1630
Coefficient of thermal expansion (C^{-1})	1.4e−005	1.2e−005	2.3e−004	1.4e−005
Specific heat (J kg^{-1} C^{-1})	780	434	296	780
Thermal conductivity (W m^{-1} C^{-1})	0.72	60.5	0.28	0.72
Resistivity (Ω m)	–	1.7e−007	–	–
Compressive ultimate strength (Pa)	4.1e+007	0	0	7e+006
Compressive yield strength (Pa)	1.5e+007	2.5e+008	0	7.5e+007
Tensile yield strength (Pa)	1.6e+006	2.5e+008	2.5e+007	2.5e+008
Tensile ultimate strength (Pa)	5e+006	4.6e+008	3.3e+007	7e+006
Temperature (C)	22			
Young's modulus (Pa)	1.89e+011	2e+011	1.1e+009	4.5e+009
Poisson's ratio	0.18	0.3	0.42	0.18
Bulk modulus (Pa)	9.84e+010	1.67e+011	2.29e+009	2.34e+009
Shear modulus (Pa)	8.01e+010	7.69e+011	3.87e+008	1.91e+009

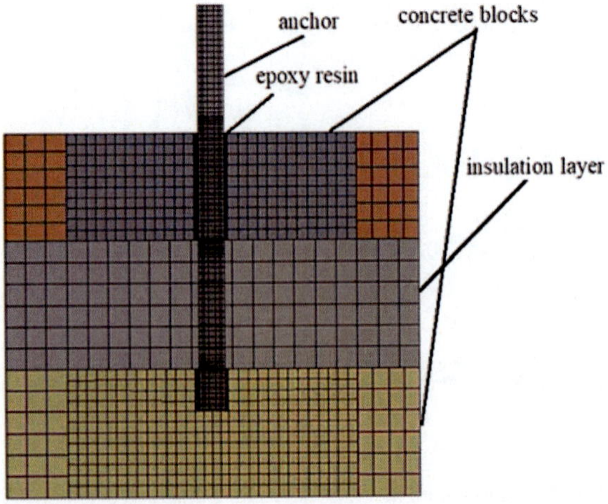

Fig. 7 Discretization by the Finite Element Method of modeled blocks and bar elements

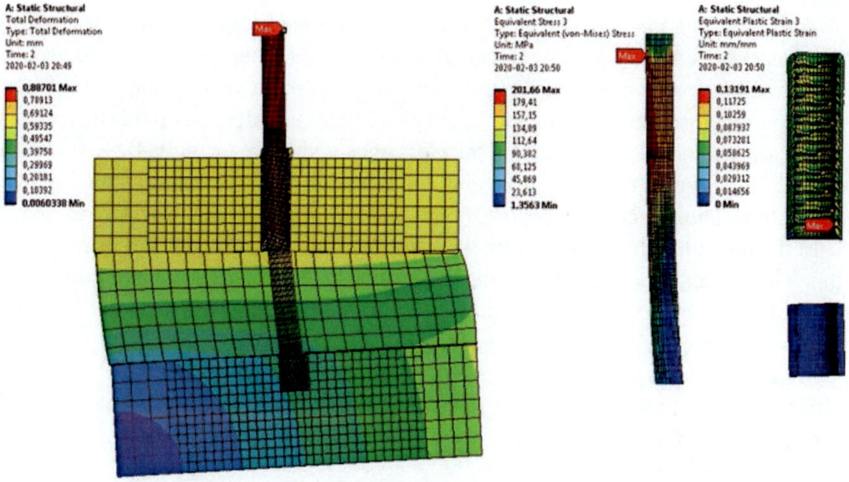

Fig. 8 View of stresses in the modeled block element, directly in the rod element (anchor) and in the contact zone of the threaded rod element with concrete block elements

concrete blocks and the concrete blocks—the insulation layer. There is an assumption that contact between the anchor and the insulation layer can be omitted as it does not influence the results. The cylindrical surface of the anchor can be divided into two areas: the non-contact area (outside the concrete blocks and inside the thermal insulation layer) and the contact area (inside the concrete blocks).

In Fig. 8 was numerically simulated the model from Fig. 7. The interaction of the pull-out force of the anchor from the concrete block element with a shear force can be observed. The shear force affects the top layer of the concrete block element by applying a constant value, i.e. 1.4 kN. In Fig. 8, we can also see the simulation of steel anchor deformation directly after loading. We can observe a contact zone simulation in the kind of a threaded steel anchor surface.

The most challenging in obtaining the convergence was the connection between anchors and concrete thought the resin material. It was solved using the non-typical approach with contact modification, which gave enough accuracy in comparison to the overall experiment. The anchor rod comes into contact with concrete through the resin filling the drilled hole with an enlarged diameter. This transition zone in numerical simulation is a solid (cylindrical ring), located between the cylindrical surfaces of the hole in the concrete and the corresponding surfaces of the anchor rod. In the numerical simulation, material parameters were applied as for the linear-plastic material (bi-linear model, with strain hardening). The parameters provided by the manufacturer were adopted, i.e. the modulus of elasticity and the tensile strength. The connection type of the cylindrical surface of the hole in the concrete with the resin solid outer surface is bonded. The anchor rod surfaces contact with the resin solid inside surfaces is realised using the modified frictional connection. This transition zone between the anchor rod and the concrete is 2 mm thick.

The modification of frictional contact mentioned above allows including the anchor thread in numerical calculations. This is the feature introduced in ANSYS 15.0. It allows to define a threaded connection between cylindrical surfaces without modelling the exact thread geometry. The most important data is necessary to provide: a mean pitch diameter, a pitch distance and a thread angle. Based on these values, the program modifies a computational model by simulating a real threaded connection. The effects of applying this modification can be observed in Fig. 9.

In the indicated drawing, regular distribution of local stresses zones draws attention. This effect is caused by the application of the contact modification described above. For concrete samples without resin, the same solution has been used to meet similar conditions as in the real experiment.

Figure 10 shows the FEM numerical simulation for a three-layer block element with an anchor at 60°. Secondary shear force was not included in this analysis, as it was not included in experimental tests (19).

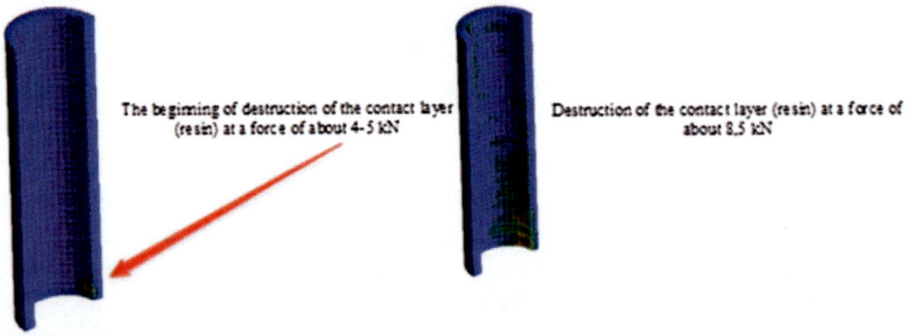

Fig. 9 View of the beginning of destruction and full cross-section destruction of the contact zone between the anchor and concrete

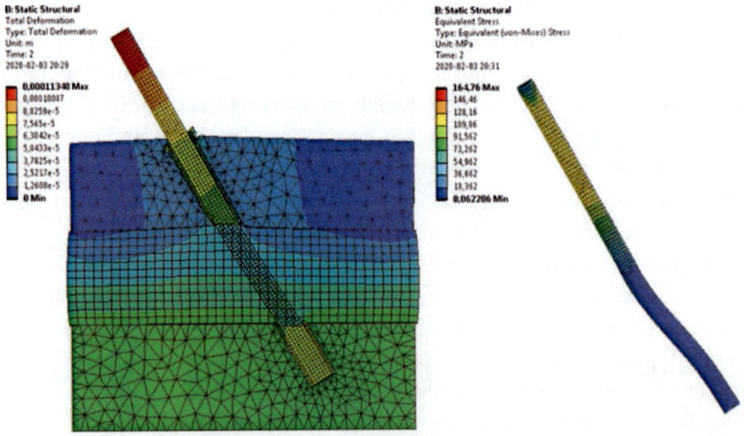

Fig. 10 A view of the numerical simulation resultant deformation of the bonded anchor in a concrete block model at an angle of 60°

Models (Figs. 11 and 12) also do not take into account the influence of the secondary shear force of the textured (upper) layer due to the limitations of the test stand. Figure 13 shows the simulation of stresses in the concrete block surface, as well as the surface of the steel anchor.

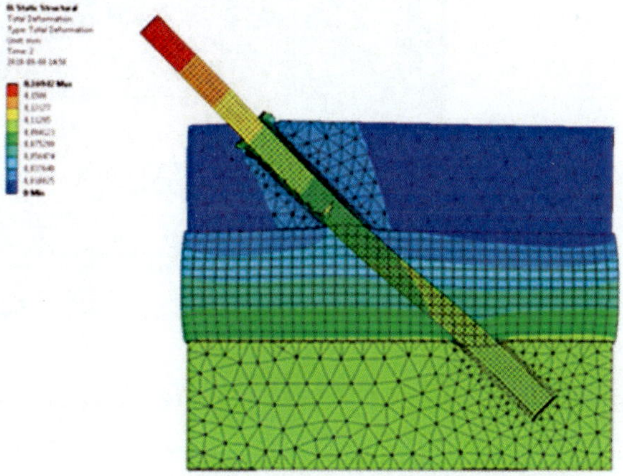

Fig. 11 A view of the numerical simulation resultant deformation of the bonded anchor in a concrete block model at an angle of 45°

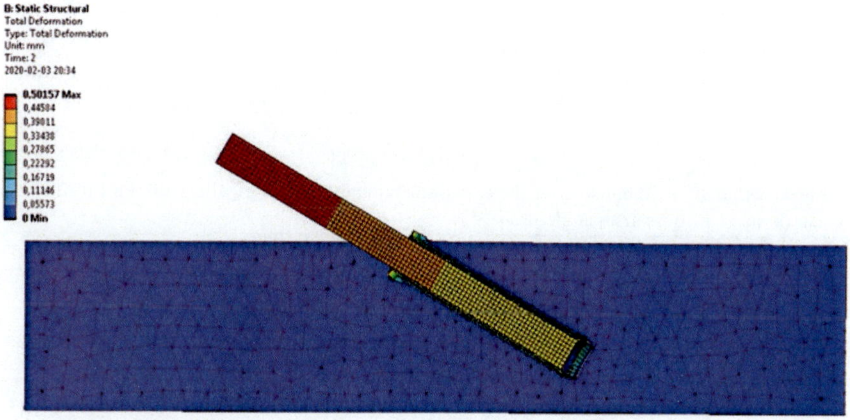

Fig. 12 A view of the numerical simulation resultant deformation of the bonded anchor in a concrete block model at an angle of 30°

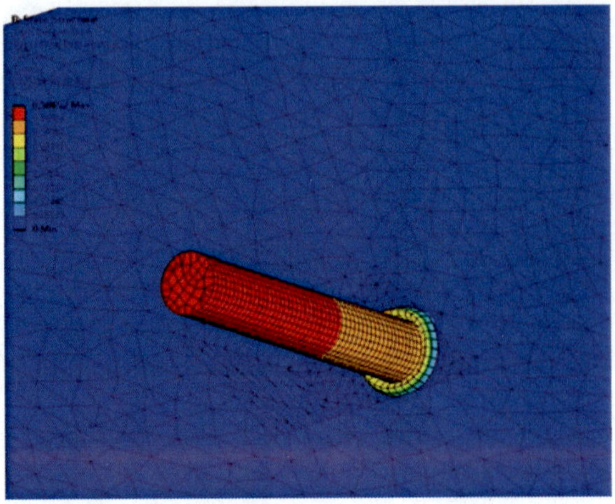

Fig. 13 View of the stress simulation on the surface of the concrete block element and steel anchor

4 Load Capacity of Bonded Anchor Tested Experimentally and Comparison to the FEM

From the carried out experimental tests, results are shown for each case: three-layer solid elements with an anchor at 90°, one at 60°, 45° and an anchor at angle 30° in the single-layer concrete block element. The results obtained in the experimental research (19) and the FEM numerical simulation of the analyzed models are summarized and compared in Tables 2, 3, 4, 5, 6, 7, 8 and 9 and in graphic form describing the force-displacement relationship in Figs. 15, 16, 17, 18, 19, 20 and 21. Preliminary simulation tests using the Finite Element Method are intended to compare with

Table 2 Summary of the results of the experimental and the numerical calculations for the first model (bonded anchors at an angle of 90°)

Description	Results	Pull-out force (kN)	Displacement (mm)
Three-layer block element with the anchor bonded 90°	Experiment	5.70	0.07
	ANSYS	8.00	0.09
Difference		2.30	0.02

Table 3 Summary of the results of the experimental and the numerical calculations for the second model (bonded anchors at an angle of 90°)

Description	Results	Pull-out force (kN)	Displacement (mm)
Three-layer block element with the anchor bonded 90°	Experiment	8.10	0.23
	ANSYS	10.40	0.10
Difference		2.30	0.13

Table 4 Summary of results of the experimental and the numerical calculations for the first model (bonded anchors at an angle of 60°)

Description	Results	Pull-out force (kN)	Displacement (mm)
Three-layer block element with the anchor bonded 60°	Experiment	11.25	0.13
	ANSYS	14.00	0.11
Difference		2.75	0.02

Table 5 Title of the inserted table summary of the results of the experimental and the numerical calculations for the second model (bonded anchors at an angle of 60°)

Description	Results	Pull-out force (kN)	Displacement (mm)
Three-layer block element with the anchor bonded 60°	Experiment	9.10	0.10
	ANSYS	14.00	0.19
Difference		4.90	0.09

Table 6 Summary of results of the experimental and the numerical calculations for the first model (bonded anchors at an angle of 45°)

Description	Results	Pull-out force (kN)	Displacement (mm)
Three-layer block element with the anchor bonded 45°	Experiment	3.50	0.04
	ANSYS	11.00	0.08
Difference		7.50	0.04

Table 7 Summary of the results of the experimental and numerical calculations for the second model (bonded anchors at an angle of 45°)

Description	Results	Pull-out force (kN)	Displacement (mm)
Three-layer block element with the anchor bonded 45°	Experiment	4.20	0.05
	ANSYS	11.00	0.09
Difference		6.80	0.04

Table 8 Summary of the results of experimental and the numerical calculations for the second model (bonded anchors at an angle of 30°)

Description	Results	Pull-out force (kN)	Displacement (mm)
Three-layer block element with the anchor bonded 30°	Experiment	11.90	0.14
	ANSYS	12.00	0.09
Difference		0.10	0.05

Table 9 Summary of results of experimental and numerical calculations for the second model (bonded anchors at an angle of 30°)

Description	Results	Pull-out force (kN)	Displacement (mm)
Three-layer block element with the anchor bonded 30°	Experiment	30.50	0.35
	ANSYS	30.00	0.33
Difference		0.50	0.02

results obtained in the experimental tests. The results from lab tests are taken as the reference.

In the case of numerical simulation, the anchor pullout force was increasing with an equal increment from 0.1 to 1 kN. To the value of 0.8 kN, the beginning of the axial vertical displacement of the anchor took place, which at this point was 0.01 mm and until the level of the pulling force equal to 1 kN, the value of the displacement of the anchor did not change. With the force increase up to 2 kN, the displacement value reached 0.02 mm. Then, in the numerical simulation, the force value stepped to 2 kN, up to a final value of 8.0 kN and a final displacement value of 0.09 mm. The FEM results correspond relatively well with experimental data (Fig. 14).

The next case is the block model with 90° anchoring. The increase in displacement occurred from the value of the force pulling 0.6 kN and then reached the value of 0.01 mm.

Then, from the value of the force of 1 kN, the value of the force increase stepped to 2.60 kN, which resulted in an increase of the displacement to the value of 0.03 mm. From the value of 2.60 kN, there was an even step of just 2.60 kN, up to a breaking force of 10.40 kN and displacement of 0.10 mm (Fig. 15). The FEM simulation is different near force of 5.00 kN.

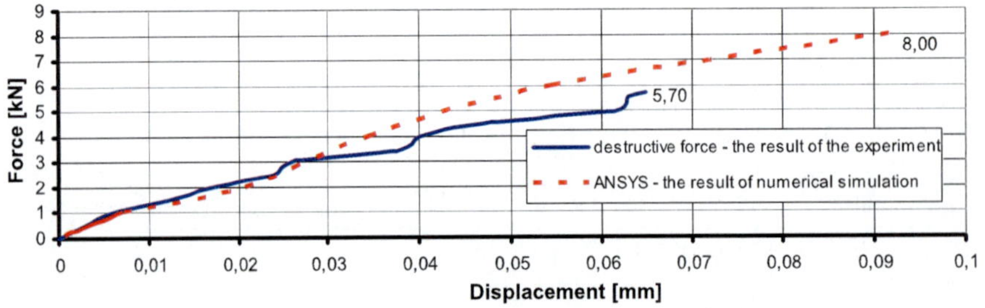

Fig. 14 Comparison of the results of the experimental tests and the numerical simulation of an anchor bonded perpendicularly to the block model surface

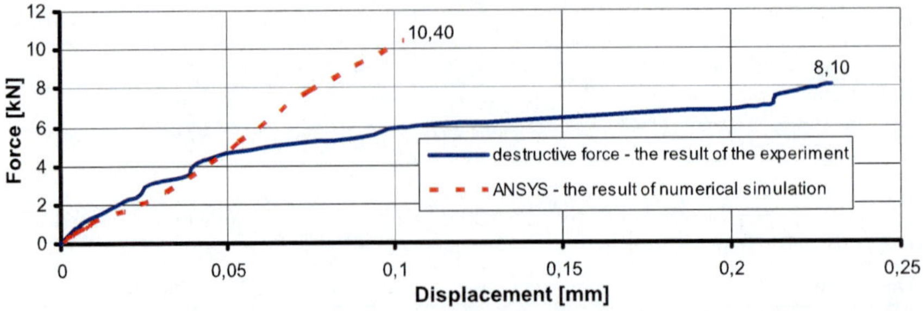

Fig. 15 Comparison of the results of the experimental tests and the numerical simulation of an anchor bonded perpendicularly to the block model surface

In the next series of the tests, the numerical analysis concerned three-layer block elements with an anchor bonded in at an angle of 60°. In this model, initially, the pull-out force increased with a 0.05 kN step, until it reached 0.5 kN. The beginning of displacement was from a force value of 0.25 kN and 0.01 mm. Another increase in the value of the pulling force is a step of 0.9 to 1.40 kN and displacement of 0.02 mm. The next steps in the pull-out force were constant until the limit value was reached and increased by 1.40 kN, up to a force of 14.00 kN and a displacement value of 0.11 mm (Fig. 16).

The comparative model of anchoring at an angle of 60° leaded to the similar approach procedure in numerical simulation, but the final stage, i.e. the destruction of the connection of the adhesive anchor with the concrete block elements, differed for a maximum pull-out force of 14.00 kN, larger final displacement that reached value 0.19 mm (Fig. 17).

Another series of tests of the load capacity of bonded anchors in block models with anchorages anchored at an angle of 45°.

In the both calculated models of this series, the stroke through the entire numerical simulation was the same until reaching the destructive force of the connection and had a value of 1 kN.

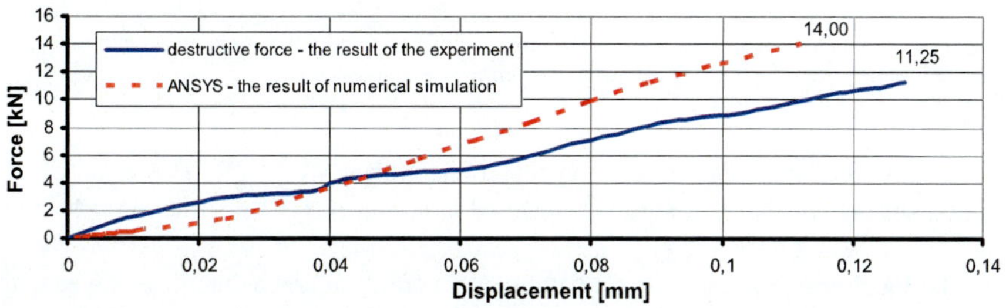

Fig. 16 Comparison of the results of the experimental tests and the numerical simulation of an anchor bonded at an angle of 60° to the surface of the block model

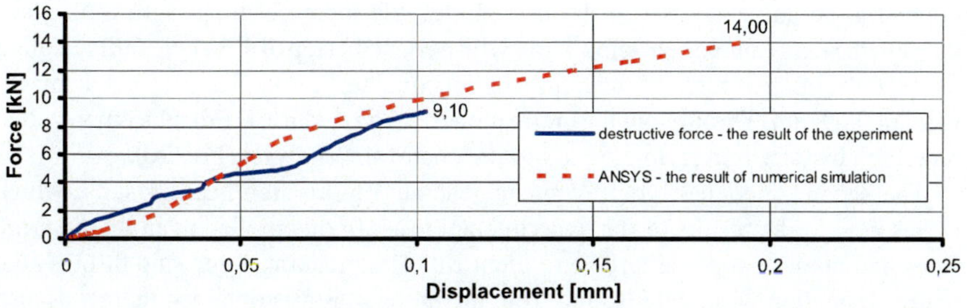

Fig. 17 Comparison of the results of the experimental tests and the numerical simulation of an anchor bonded at an angle of 60° to the surface of the block model

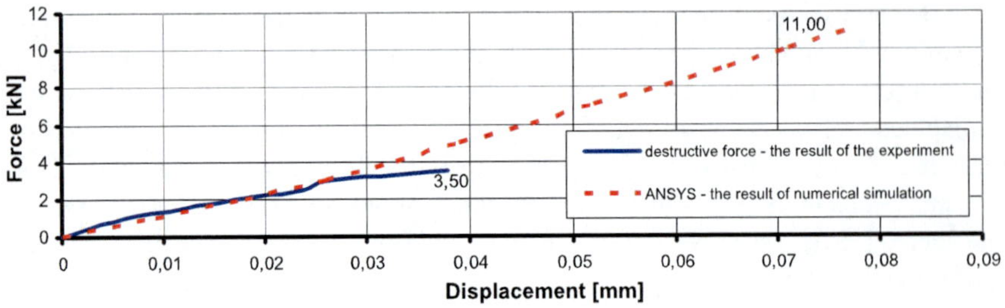

Fig. 18 Comparison of the results of the experimental tests and the numerical simulation of an anchor bonded at an angle of 45° to the surface of the block model

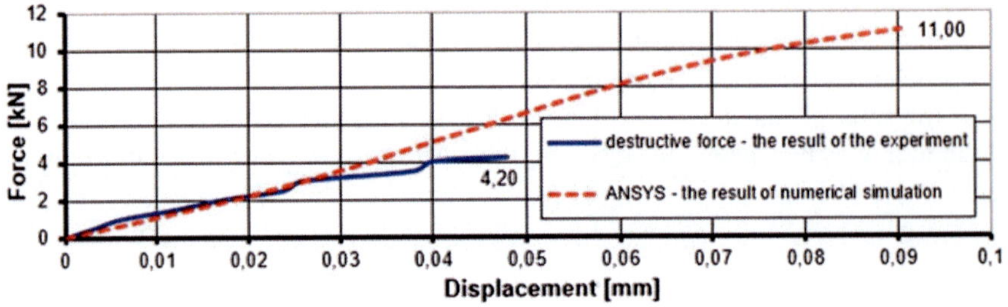

Fig. 19 Comparison of the results of the experimental tests and the numerical simulation of an anchor bonded at an angle of 45° to the surface of the block model

The value of the force destroying the connection is 11.00 kN. The value of the final displacement in the first model reached 0.08 mm (Fig. 18), while in the second model, it reached 0.09 mm (Fig. 19).

The block models of the first six elements were designed as the three-layer with dimensions of 20 cm × 20 cm × 17 cm. These are respectively: a 5 cm thick texture layer, a 6 cm thick foam polystyrene layer, and a 6 cm thick construction layer.

The last two block models were designed as a single-layer with dimensions of 30 cm × 20 cm × 6 cm due to the sharp angle of inclination (30°).

The last series of block models concerned single-layer models with an anchor bonded at an angle of 30°. In the first model, the increase in strength took place through the entire numerical simulation with a constant step of 1.50 kN, until reaching the final destructive force of the bonded joint equal to 12.00 kN and a final displacement of 0.09 mm. Throughout the entire numerical procedure, the displacement value increased by 0.01 mm from 1.50 kN to the end of the analysis (Fig. 20).

The second modeled block element generated more significant results, which correspond to the results of the experimental tests. In this model, as in the first, the steps and numerical procedure were identical. The breaking force step through the entire simulation was 1.50 kN, and from the moment it was obtained there was also a displacement that reached its apogee with a breaking force of 30.00 kN and had a value of 0.33 mm (Fig. 21).

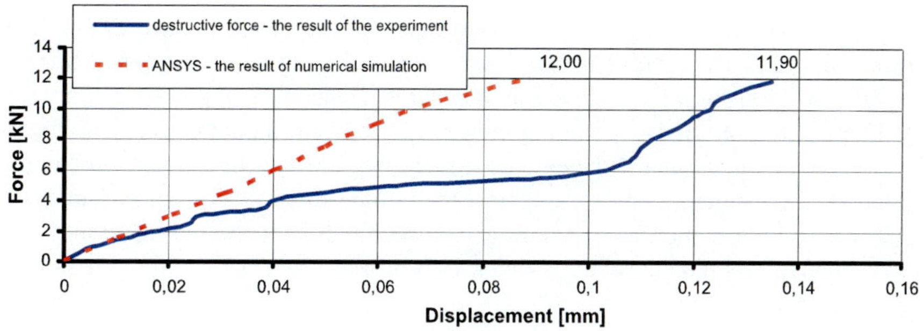

Fig. 20 Comparison of the results of the experimental tests and numerical simulation of an anchor bonded at an angle of 30° to the surface of the block model

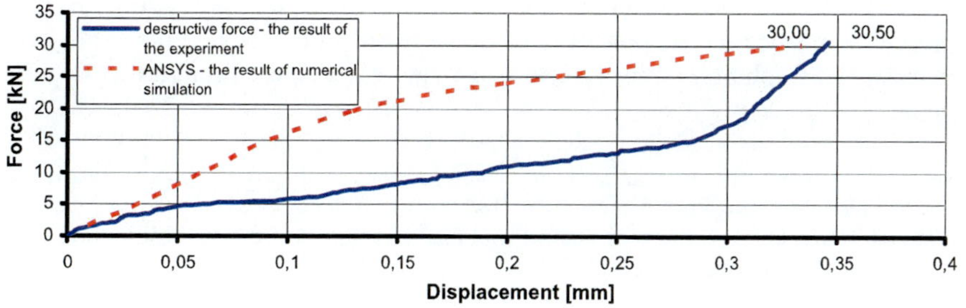

Fig. 21 Comparison of the results of the experimental tests and numerical simulation of an anchor bonded at an angle of 30° to the surface of the block model

5 Conclusions

The complex set of experimental tests has been carried to check the behavior and capacity of multilayered wall connections. The research was conducted for the purpose of traditional large-panel residential buildings where the connections are the most challenging in case of the definition of their strength. The tests were finished with the preliminary simplified numerical calculations, which were useful as the partial validation of the tests and showed potential future fields of studies. Analyzing the obtained numerical simulation and comparing them with experimental research, it can be concluded that the computed results are close in some cases and relatively large in others. These differences can be caused by numerical model simplifications. Cracking and cruising effects in concrete were not taken into account due to possible computing consuming high nonlinearities, which could appear during calculations for all the cases. The main aim was to build an overview estimation as the initial step. A more detailed simulation would solve the differences in the results for the chosen types of connections basing on calculations described here.

The most considerable differences were identified in the case of three-layer block models with a 45° anchorage model. Differences appeared for cases with forces ranged from 0.10 to 7.50 kN with measured displacements from 0.02 to 0.13 mm. The differences may also result from the preparation accuracy of models for laboratory experiments, i.e. during the formation of concrete elements or application of a resin. Random cracks of relatively small dimensions or other secondary effect and imperfections could appear in solid elements, e.g. during the holes drilling process, could influence the resultant resistance against the pulling force. Such types of analyzes and comparative numerical simulations can be helpful for the real assessment of repairs and reinforcements of three-layer walls in large slab panel buildings in Poland [20, 21].

It worth to be mentioned that in the simulations performed in ANSYS, the threaded part of a bonded anchor has been taken into account. It has been done by modification of standard "frictional" contact type by applying the "virtual thread" conditions. This software feature is predicted mainly to be used in conventional bolted connections (e.g. between steel plates) to reduce computational costs. In this paper, it has been chosen to enhance the accuracy of contact between steel threaded bars with resin material. This solution is not typical and shall be considered for further detailed studies.

Acknowledgements The research was carried out with the use of software ANSYS 19.2, which was provided to the Bialystok University of Technology on the basis of an agreement between Bialystok University of Technology and ANSYS Inc. (Canonsburg, USA) and MESco Sp. z o.o. (Tarnowskie Góry, Poland).

References

1. Pukl R, Ožbolt J, Eligehausen R (1998) Load-bearing behavior of bonded anchors—nonlinear simulation. In: Computational modelling of concrete structures, Rotterdam
2. Li Y, Eligehausen R, Ožbolt J (1999) Numerical analysis of quadruple fastenings with bonded anchors
3. Eligehausen R, Appl J (2007) Behavior and design of fastening with bonded anchors: Numerical analysis and experimental verification. In: VI international conference on fracture mechanics and concrete structure
4. Bajer M, Barnat J (2012) The glue–concrete interface of bonded anchors. Constr Build Mater 34:267–274
5. Sonoda Y (2016) A numerical study on the pull-out strengths of anchor bolts embedded in concrete using the smoothed particle hydrodynamics method. Key Eng Mater 711:1111–1117. ISSN 1662-9795
6. Prieto-Muñoz PA, Yin HM, Testa RB (2010) An elastic analysis that predicts the pull-out capacity of adhesive anchors. Mater Sci Eng 10:1–10. WCCM/APCOM
7. Spada A, Giambanco G, Rizzo P (2011) Elastoplastic damaging model for adhesive anchor systems. I: theoretical formulation and numerical implementation. J Eng Mech 854–861
8. Cruz-Sena J, Cunha VMCF, Camoes A, Barros JAO, Cruz P (2009) Modelling of bond between galvanized steel Rebars and concrete. In: Congreso de Metodos Numericos en Ingenieria 2009, Barcelona, 29 junio al 2 de junio 2009, SEMNI, Spain, pp 1–15
9. Ningyu Z, Junbo Z, Haifei (2019). Analysis on the tensile strength of post-installed large-diameter anchors in concrete. J Eng Sci Technol Rev 12(5):112–121

10. Serrano E (2001) Glued-in rods for timber structures—a 3D model and finite element parameter studies. Int J Adhesion Adhesives 21:115–127
11. Szczecina M, Winnicki A (2016) Selected aspects of computer modeling of reinforced concrete structures. Archives of Civil Engineering, vol. LXII(1)
12. Nuralinah D (2019) The pull-out test on knit bamboo reinforcement embedded into concrete beam. MATEC Web Conf 258:1–7
13. Hariyadi MS, Sonoda Y (2017) Experimental analysis of anchor bolt in concrete under the pull-out loading. Procedia Eng 171:26–933
14. Bažant ZP, Şener S (1988) Size effect in pullout tests. ACI Mater J 347–351. Technical Paper. Title No. 85-M38
15. Epackachi S, Esmaili O, Mirghadeli SR, Behbahani AAT (2015) Behavior of adhesive bonded anchors under tension and shear loads. J Constr Steel Res 114:269–280
16. Hüer T, Eligehausen R (2007) Splitting failure mode of bonded anchors. In: Fracture mechanics of concrete structures—design, assessment and retrofitting of RC structures, pp 753–760
17. Kijania M (2015) The methods for computation of the bond stress between concrete and steel reinforcement. Przeglad Budowlany 38–42 (in Polish)
18. Ligocki I (2002) HYSDOZOK—an automated system for applying loads in structural tests. Konstrukcje Stalowe (3) (in Polish)
19. Hyrczak E, Kolarzowski M (2003) The HYSDOZOK hydraulic system is used to apply structure loads. Hydraulika i Pneumatyka 4:17–18 (in Polish)
20. Szlendak JK, Jablonska-Krysiewicz A, Tomaszewicz D (2018) Assessment of the load capacity of the anchorage system connecting the textured layer with the structural wall of large slab buildings in the lights of experimental research and FEM analysis. In: ECCE Opole
21. Szlendak JK, Jablonska-Krysiewicz A, Tomaszewicz D (2018) Comparative analysis of oblique bonded anchors with point anchors fixed in the concrete structural layer of buildings of large slab. In: 3rd world multidisciplinary civil engineering—architecture—urban planning symposium; WMCAUS 18–22 June 2018, At Prague, Czech Republic

Sustainability of Concrete Structures in Terms of Concrete Frost Resistance Determination

D. Kocáb, P. Daněk, P. Žítt, and T. Vymazal

Abstract This chapter deals with determining the frost resistance of concrete built in a structure. The test result can be used as one of the input parameters in sustainability indicator calculation, e.g., when there is a need to assess the suitability of different concrete structure repair designs in terms of sustainability. Almost all standards dealing with concrete frost resistance prescribe testing specimens made in moulds. This is a fundamental problem when testing an existing concrete structure. The aim of the experiment described is to propose a method for determining the frost resistance of concrete by testing core samples. Several options were exercised. The non-destructive methods include the ultrasonic pulse velocity and resonance methods, while the destructive method includes the tensile splitting strength test.

Keywords Sustainability · Concrete · Frost resistance · Concrete structure

1 Introduction

In 2015, the European Commission adopted the Community Programme LIFE, whose primary goal is to contribute to a significant development of low-emission economy. Such economy would use all sources effectively, it would be environmentally friendly, it would take into consideration the effort to mitigate climate change, and it would contribute to protecting and improving the environment, the countryside, and biodiversity [1–3]. On that account, sustainability has been a phenomenon in the recent years and it must be said that rightfully so. It is an approach that represents the future of all activities, including construction. Not only should the design be assessed with regard to sustainability but also the process of construction, operation, maintenance, and repairs of all structures—including the concrete ones [4].

Concrete is the most common building material all over the world. Therefore, it represents a considerable burden to the environment, especially in terms of CO_2

D. Kocáb (✉) · P. Daněk · P. Žítt · T. Vymazal
Faculty of Civil Engineering, Brno University of Technology, Veveří 331/95, 602 00 Brno, Czech Republic
e-mail: dalibor.kocab@vutbr.cz

Z. Zembaty et al. (eds.), *Environmental Challenges in Civil Engineering*, Lecture Notes in Civil Engineering 122,
https://doi.org/10.1007/978-3-030-63879-5_5

emissions. Despite that, there is no doubt it will be used as the primary building material in the near future because it has a great number of advantages (e.g. high variability of its composition and properties and the shape variability of the elements made from it) [5]. It is therefore obvious that there is great scope for searching for and developing new technologies and procedures when designing new concrete structures or assessing the current ones, including their assessment with regard to sustainability.

General principles related to sustainability in construction are described in the ISO 15392 standard [6]. The purpose of the ISO 13315 standard [7, 8] is to provide the basic rules of environmental management for concrete and structures built from it. The primary goal is to help mitigate the effects on the environment resulting from activities related to construction of concrete structures. The point of sustainable design of structures is to reduce the negative impact and, at the same time, increase the benefits for the society, the environment, and economy [4, 9].

Although much has been written on sustainability recently and at first sight it might appear to be a simple matter, it is in fact a rather complicated issue [10]. There are several definitions that may differ based on the area of the construction industry being assessed. The basis of sustainability is formed by three primary aspects (pillars): social, environmental, and economic. Two areas of defining sustainability can be used: general and construction-related. The first definition (the general one) characterizes sustainability in a rather broad context, not being linked to a specific area. The other definition (the construction-related one) focuses on more specific situations, e.g. sustainability of building structures [6]. In relation to the implementation of the sustainable development principles there is a very important question of whether the social, economic, and environmental aspects should be given equal importance or whether environmental protection is more important and therefore the environmental aspect of sustainability is above the others. Regardless of more and more frequent discussions on the significance of each pillar, it is generally recognized that all the three pillars must be taken into consideration in relation to sustainable development [4]. However, it is certainly necessary to consider sustainability from the very start of the construction process.

2 Sustainability of Concrete Structures with Regard to Frost Resistance

As is clear from the introduction, sustainability of concrete structures is a very broad topic whose practical application is complicated. If it is necessary to decide about the suitability of various design options of concrete structures in compliance with the principles of sustainability, a clearly defined criterion must be available [11]. One such criterion could be the building material sustainability potential (BMSP). It is a rather simple way of comparing the sustainability levels of various kinds of concrete with regard to their resistance to various degradation effects. The BMSP assessment criterion is based on [12, 13], where the three pillars of sustainability are described with respect to three variables describing concrete—performance, service life, and environmental impact:

$$BMSP = \frac{performance \times servicelife}{environmentalimpact} = \frac{R \times L}{E} \tag{1}$$

The building material sustainability potential can subsequently be transformed into what is called a sustainability indicator k_{SB} [4]. It is a coefficient created by normalizing Eq. (1) and can be used to quantify sustainability in relation to the surrounding conditions in which the concrete is found and to the expected development of its degradation:

$$k_{SB} = \frac{\frac{R}{R_{ref}} \cdot \frac{L}{L_{ref}}}{\frac{E}{E_{ref}}} \tag{2}$$

The variable R represents performance, L stands for service life (or another service life indicator), and E represents costs. All these values, describing the concrete variety, are divided by suitably selected reference values of L_{ref}, R_{ref} and E_{ref}. The result is a dimensionless variable k_{SB}, whose value ranges around 1. Performance R may represent (most often) compressive strength or any other property of concrete—e.g. frost resistance. The variable E (eco-costs) represents the money needed for measures to be taken to reduce the environmental impact to a sustainable level; see more e.g. in [4, 14]. However, there are many other definitions of the variable E such as global warming potential, emission allowance price, carbon footprint, and others. The calculation may also include price:

$$k_{SB,C} = \frac{\frac{R}{R_{ref}} \cdot \frac{L}{L_{ref}}}{\frac{E}{E_{ref}} \cdot \frac{C}{C_{ref}}} \tag{3}$$

However, the cost of the material C depends, to a certain extent, on the location where the concrete structure is being designed or built. Therefore, a general evaluation which could be applied worldwide cannot be made. The above-mentioned procedure can also be carried out using a fully probabilistic approach [4].

When assessing, with regard to sustainability, different concrete designs intended for construction of a new concrete structure exposed to rain and frost (and frost resistance is one of the predominant properties of concrete), the procedure is quite easy. In the calculation of k_{SB} (or $k_{SB,C}$), the frost resistance parameter of each concrete is substituted for R. The other variables from Eq. (2), or Eq. (3), depend on the composition of each concrete and the resulting sustainability indicator determines the most suitable concrete. When entering the parameter R, we must make sure that it is identical for all the concrete designs assessed—i.e. the frost resistance parameter being uniquely determined by the number of equal freezing and thawing (F-T) cycles. It could, for example, be the relative dynamic modulus of elasticity P_c or the length change L_c in accordance with the American standard of ASTM C666/C666M—15 [15], the relative dynamic modulus of elasticity RDM_n or the relative length change $\varepsilon_{L,n}$ in accordance with one of the procedures described in the European standard of CEN/TR 15177 [16] or the frost resistance coefficient in accordance with the Czech standard of ČSN 73 1322 [17], which is the ratio of the flexural strength of the test specimens after F-T cycles to the strength of the reference specimens (which were not burdened with the F-T cycles). Another European document of RILEM TC 176-IDC [18] is almost identical to CEN/TR 15177 [16].

If there was a need to assess a concrete structure repair project (using the indicator k_{SB}) with regard to sustainability, the situation would be significantly more complicated, namely for two reasons. For one thing, it would be rather problematic to assess different repair options using k_{SB} because there are other aspects as well than just comparing the different concrete recipes; for another, it is complicated to determine the frost resistance of concrete built in a structure. The following text deals with the latter of the problems described.

3 Determining the Frost Resistance of Concrete

If concrete in a structure is exposed to repeated alternation of positive and negative temperatures, its degradation occurs if it is also exposed to water or a high level of moisture. If dry concrete is frozen and thawed in an environment with low air humidity, no considerable negative impact is created [19]. With newly designed concrete it is not difficult to check frost resistance—appropriate tests are done on the test specimens, which results in determining the resistance level of the concrete to freezing and thawing. However, if it is necessary to check this parameter in older concrete built in a structure (e.g. during reconstruction or structural repairs), it often poses a fundamental problem.

If the concrete structure or its part is in contact with water (road panels, railway sleepers, water tanks etc.), the water gets into the porous structure of the concrete due to its absorbability. If it freezes, there is a risk of micro cracks and later even cracks occurring in the concrete. It is caused by water expansion as it changes from liquid to solid state. Water present in the capillary pores in concrete starts to change into crushed ice at the temperature of approximately −0.5 °C (the point when the water starts freezing depends on the size of the pores; however, all the capillary water should freeze at −12 °C) [20]. The volume of the ice formed is approximately 9% larger than that of water in the liquid state. This water expansion in the form of ice leads to internal stress development in the concrete, namely of the size of tens of MPa. This results in the disruption of the inner concrete structure and subsequent irreversible reduction in its fundamental material properties [21].

There are several ways of determining the frost resistance of concrete. There is a limitation consisting in the fact that the testing is done almost solely on test specimens made in moulds. The standard [15] is an exception allowing the testing of drilled core specimens or prisms cut out of hardened concrete. In Europe, however, only specimens from moulds are always tested. The principle of frost resistance determination is monitoring the decrease in a certain concrete property value after running a certain number of F-T cycles with the presence of water. The Czech standard [17] works primarily with flexural strength, which is logical—as soon as cracks or even micro cracks appear in the inner concrete structure, the flexural strength immediately decreases. The standards [15, 16] determine the frost resistance of concrete by monitoring changes in its dynamic modulus of elasticity or changes in the length of the specimens after F-T cycles. Testing the concrete dynamic modulus of elasticity is also suitable for this purpose as it is sensitive to changes in its inner structure quality. The result is the relative dynamic modulus of elasticity (further marked as RDM).

4 The Experiment

As is clear from the previous text, the aim of the experiment was to determine the frost resistance of concrete already built in a structure. It must be noted that the frost resistance of concrete was tested without using chemical defrosting agents (some of the above-mentioned procedures [16, 18] prescribe that concrete should be tested using a sodium chloride solution).

4.1 Test Specimens and Material

As has already been mentioned, the European standards require that the testing should be done using test specimens made in moulds, preferably prisms of 100 × 100 × 400 mm. The document [16] does not set a requirement for the number of test specimens. The standard [17] says that the number of test specimens depends on the number of the F-T cycles planned. Generally, it can be stated that a suitable number of test specimens that should be available is 9. When determining the frost resistance of concrete in a structure, a problem usually already occurs during the sampling of the specimens. It is very difficult to cut out a prism with the lateral dimensions of 100 × 100 mm of a standard core sample with a diameter of 150 mm—either the cut out prism will have rounded edges or its lateral dimensions will be 95 × 95 mm at best. For the test to be completely correct according to the standards, core samples with a diameter of 160 mm or preferably 180 mm would have to be taken from the structure. Moreover, the length of such a (undisturbed) core sample would have to be almost 0.5 m (or its multiple) and the total length of the core samples (arranged one after another) would have to be approx. 4.5 m. In addition, the core samples could not be reinforced in the place where the 100 × 100 × 400 mm prism would later be cut out, which means in its centre. It is very difficult, in some cases even impossible, to obtain the test specimens required for frost resistance tests from almost any existing concrete structure.

Core samples (further on also CS) from reinforced road bridge abutments were used for the frost resistance test. Due to the conditions imposed by the bridge manager, it was not possible to extract a CS with a diameter of over 100 mm. It was therefore clear that the concrete frost resistance will not be determined using standard prism specimens of 100 × 100 × 400 mm. CS with a nominal diameter of 100 mm were taken from both abutments and test cylinders with a nominal diameter of 100 mm and a nominal length of 150 mm were made from them. These test specimens were used to determine the frost resistance. We also prepared test specimens with a nominal diameter of 100 mm and a nominal length of 100 mm for testing compressive strength and specimens with a nominal diameter of 100 mm and a nominal length of 200 mm for testing direct tensile strength.

To determine the frost resistance, the original plan was to use 9 test cylinders, 3 of which were supposed to be reference specimens and 6 were going to be exposed to the burden of F-T cycles. Three of the burdened specimens were going to be used for a destructive method of frost resistance determination after running 25 F-T cycles and the remaining three were meant to be used for a destructive method of frost resistance determination after 50 F-T cycles. However, it turned out that a test planned in this way could not be done with a relevant result since each of the abutments was very probably made from a different kind of concrete, see Fig. 1. That is why the number of test specimens was increased to 10 and 5 test cylinders were made from each concrete—1 as a reference specimen, 2 for the test after 25 F-T cycles, and 2 for the test after 50 F-T cycles. This number of test cylinders made it possible, albeit in a limited way, but still, to assess each concrete separately as a statistical data set as well as both concretes together as one set. The first concrete was marked with the letter "A" and the other concrete with the letter "A". Concrete B was apparently made with a smaller amount of coarse aggregate (Fig. 1). The basic parameters defining the concretes are shown in Table 1.

Fig. 1 Test cylinder from a core sample of concrete A on the left; from concrete B on the right

Table 1 Average values and coefficients of variation (CoV) of the basic properties determined for concretes A and B

Property	Concrete A		Concrete B	
	Average	CoV (%)	Average	CoV (%)
Density (kg/m^3)	2270	0.85	2260	0.82
Compressive strength (MPa)	26.0	4.85	24.9	4.28
Direct tensile strength (MPa)	1.4	25.56	1.1	19.26

4.2 Testing Methods

At the beginning of the experiment it was decided that the concrete will be tested in accordance with the European standards. The process of freezing and thawing took place according to the Czech ČSN 73 1322 standard [17], primarily because the procedure is more simple. One freezing and thawing cycle consists of 4 h of freezing, when the air temperature ranges from −15 to −20 °C, and 2 h of thawing in water with a temperature of +20 °C. So one F-T cycle takes 6 h. Cooling to the required temperature is continuous and takes 1.5 h. The test specimens undergo the required number of cycles in stages, usually stages of 25 cycles, i.e. after 1 week. It is a considerably faster cycling than what is described in [16] (and it is also easier to carry out).

With the above-described test specimens it is not possible to test the standard four-point bending as with prisms and therefore tensile splitting strength was selected as the assessment criterion instead of flexural strength. It is of course possible to determine this strength in cylinders (or core samples). The EN 206 + A1 [22] standard even prefers testing the tensile splitting strength to flexural strength. In addition to testing the concrete frost resistance using the tensile splitting strength, RDM was also determined using both the ultrasonic pulse velocity method and the resonance method.

First, the dimensions and weight of all the test cylinders were determined and then non-destructive measurements were conducted—in accordance with the EN 12504-4 [23] standard the ultrasonic pulse velocity (UPV) in concrete V was determined. The Pundit PL-200 device was used for the measurements with 150 kHz frequency probes (Fig. 2). In each test specimen, the ultrasonic transit time was determined three times in the longitudinal axis. Based on the ČSN 73 1372 [24] standard, the longitudinal vibration frequency f_L was then determined. Each test specimen placed on a flexible mat was set oscillating by a mechanical impulse using an impact hammer and the frequency was established using an acoustic emission sensor connected to the Handyscope HS4 oscilloscope and using software based on the principle of fast Fourier transform (Fig. 3).

After the non-destructive measurement, tensile splitting strength f_{ct} was determined, in accordance with the EN 12390-6 [25] standard, for the reference specimens—one test cylinder from concrete A and one from concrete B, see Fig. 4. Then

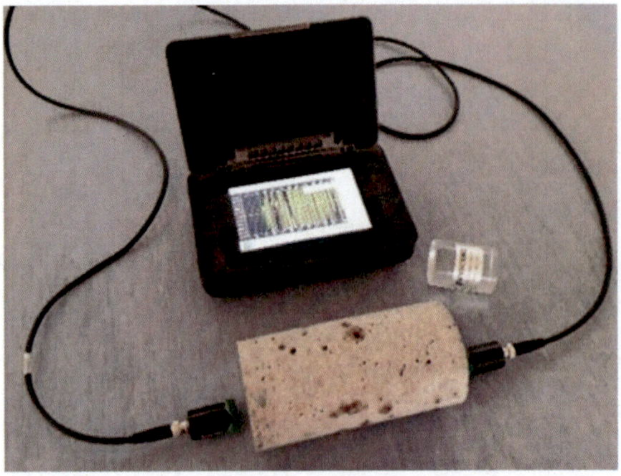

Fig. 2 Determining the ultrasonic pulse velocity in concrete

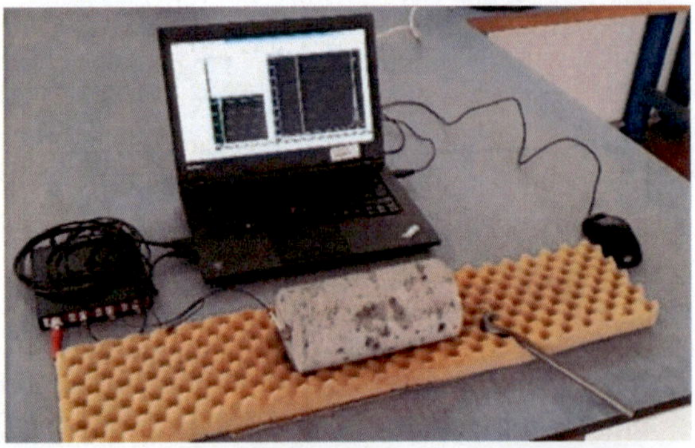

Fig. 3 Determining the resonance frequencies of the specimens

Fig. 4 Determining the tensile splitting strength of the test cylinder from concrete A

the remaining 8 test cylinders (4 from each concrete) were placed in the KD-20 automatic freezing box, where they were subjected to 25 F-T cycles described above. After the specimens were removed from the freezing box, V and f_L were determined again for all the specimens and tensile splitting strength for half of them (2 specimens from each concrete). The remaining cylinders were subjected to another 25 F-T cycles, after which the monitored parameters were determined again.

5 Results and Discussion

The values obtained for the ultrasonic pulse velocity V were used to calculate the relative dynamic modulus of elasticity RDM(U) and the determined natural longitudinal vibration frequencies f_L were used to calculate the relative dynamic modulus of elasticity RDM(FL). The calculation of RDM was based on [16]. All the results related to the NDT parameters are shown in Tables 2 and 3. The average RDM values were calculated as the ratios of the squares of average values of V (or f_L) after F-T cycles to those before F-T cycles—not as a simple average of the RDM values (determined for each test specimen). This calculation method was selected to correspond with the calculation of the frost resistance coefficient established by the strength tests.

The frost resistance coefficient of concrete after 25 and 50 F-T cycles was established based on the tensile splitting strengths, see Table 4. According to the standard [17], the frost resistance coefficient is calculated as a ratio of the average flexural strength of the frozen specimens to the average tensile strength of the non-frozen

Table 2 Velocity V values and RDM (U) values calculated from V after F-T cycles (c.)

Specimen	UPV V (km/s)			RDM (U) (%)		
	0 c.	25 c.	50 c.	0 c.	25 c.	50 c.
A1	4.394	3.435	–	100.0	61.1	–
A2	4.371	4.051	3.541	100.0	85.9	65.6
A3	4.447	3.806	–	100.0	73.3	–
A4	4.435	–	–	100.0	–	–
A5	4.479	4.238	3.516	100.0	89.5	61.6
B1	4.214	3.721	–	100.0	78.0	–
B2	4.212	4.062	4.006	100.0	93.0	90.5
B3	4.147	–	–	100.0	–	–
B4	4.362	4.185	4.096	100.0	92.0	88.2
B5	4.084	3.887	–	100.0	90.6	–
Average—concrete A					77.0	63.9
Average—concrete B					88.9	92.9
Average—both concretes					82.7	77.2

Table 3 Frequency f_L values and RDM(FL) values calculated from f_L after F-T cycles (c.)

Specimen	Natural frequency f_L (Hz)			RDM (FL) (%)		
	0 c.	25 c.	50 c.	0 c.	25 c.	50 c.
A1	12,418	8200	–	100.0	43.6	–
A2	12,468	10,753	8189	100.0	74.4	43.1
A3	12,804	9288	–	100.0	52.6	–
A4	12,485	–	–	100.0	–	–
A5	12,669	11,564	8688	100.0	83.3	47.0
B1	11,731	10,355	–	100.0	77.9	–
B2	11,462	10,995	10,867	100.0	92.0	89.9
B3	10,557	–	–	100.0	–	–
B4	12,150	11,566	11,254	100.0	90.6	85.8
B5	11,345	10,891	–	100.0	92.2	–
Average—concrete A					62.7	45.1
Average—concrete B					91.5	93.3
Average—both concretes					75.7	65.9

Table 4 Results for the tensile splitting strength f_{ct} and the frost resistance coefficients after F-T cycles (c.)

Specimen	Tensile splitting strength f_{ct} (N/mm^2)			Frost resistance coefficient (%)		
	0 c.	25 c.	50 c.	0 c.	25 c.	50 c.
A1	–	2.14	–	–	63.4	–
A2	–	–	2.18	–	–	64.7
A3	–	3.56	–	–	105.4	–
A4	3.37	–	–	100.0	–	–
A5	–	–	2.51	–	–	74.4
B1	–	1.25	–	–	58.8	–
B2	–	–	1.54	–	–	72.3
B3	2.13	–	–	100.0	–	–
B4	–	–	2.01	–	–	94.3
B5	–	1.95	–	–	91.5	–
Average—concrete A					84.4	69.6
Average—concrete B					75.1	83.3
Average—both concretes					80.8	74.9

(reference) specimens. In this calculation, the flexural strength was replaced with the tensile splitting strength—therefore, the frost resistance coefficient is not determined exactly by the standard [17] (which has already been noted in the text above); however, the meaning and principle of the calculation is precisely the same.

If it is necessary to decide whether a concrete is or is not frost resistant (of course, in relation to a certain number of F-T cycles), a minimal value must be defined below which the frost resistance coefficient (or RDM) must not drop. The limit value given by the Czech standard [17] is 75%, whereas the American standard [15] prescribes to finish the test if the relative dynamic modulus falls below 60%. The European regulations [16, 18] do not mention any criterion at all. For that reason, 75% is regarded as the limit value even for the NDT measurements in the assessment of the experiment described. Tables 2, 3 and 4 show the average RDM values and the frost resistance coefficients for each concrete separately and for both concretes together (as for one statistical data set). Based on the tests, it can be concluded that concretes A and B are probably frost resistant for 25 freezing and thawing cycles within the meaning of the ČSN 73 1322 [17] standard. However, it must be noted that this evaluation is not completely unequivocal. Concrete A, if assessed separately, is not satisfactory for 50 F-T cycles according to any of the methods used and it is satisfactory in terms of tensile splitting strength and UPV for 25 F-T cycles. Based on the resonance method, it is not satisfactory even for 25 F-T cycles, which is caused by considerably worse results of specimen A1 compared to the other test cylinders. Concrete B, again if assessed separately, is satisfactory based on the average values of all the test methods used for both 25 and 50 F-T cycles. However, the assessment must take into consideration that after 25 F-T cycles one cylinder (B1) out of two was not satisfactory in terms of tensile splitting strength and after 50 F-T cycles one cylinder (B2) out of two was not satisfactory in terms of tensile splitting strength. Of course it is a numerically completely unsatisfactory statistical set. However, the fact that half of the specimens are not satisfactory must be taken into account. When both concretes are assessed together (which is logical in terms of assessing the structure as a whole but not suitable in terms of combining two statistical data sets—meaning two concretes with different composition) the concrete is frost resistant based on all the methods for 25 F-T cycles; based on the resonance method and the tensile splitting strength it is not frost resistant for 50 F-T cycles.

Clearly, if specimens are obtained by core sampling and subsequent circular cutting, the results can be different from those obtained by testing specimens from a mould. This is true even for the same concrete being tested. The standard test specimens mature in prescribed conditions (mostly in a water bath), which cannot be said about concrete in a structure. When extracting and adapting a test specimen from a structure, micro cracks may occur, which might have a negative impact on its frost resistance and, moreover, the specimens tested have an "open" surface. According to the standard [26], if core samples are used to assess the concrete compressive strength, it is sufficient for the result to reach 85% of the characteristic value of concrete compressive strength. Concerning the assessment of frost resistance, it would be appropriate to admit a similar tolerance as well. However, it should also be noted that the drops in the flexural strength and the tensile splitting strength after the F-T cycles are not the same—the drop in the tensile splitting strength tends to be lower, see [27, 28]. Thus, assessing the concrete frost resistance is not so strict, which could compensate for the above negative impacts (thus substituting the tolerance in the sense of the standard [26]).

When assessing a concrete structure repair project in terms of sustainability with frost resistance being among the parameters used for calculating the sustainability indicator k_{SB}, the assessment criterion (i.e. the limit value for deciding whether a concrete is frost resistant) is virtually superfluous. Important is that the frost resistance parameter substituted into the k_{SB} calculation should be obtained by an identical method regardless of whether it is RDM by [16] after 100 F-T cycles or, e.g., the frost resistance coefficient by [17] after 50 F-T cycles. The only problem is, when assessing the suitability of each concrete structure repair project in terms of sustainability, how to deal with the relevant calculation of k_{SB} (thus, with the input values other than the frost resistance parameter) or of some other sustainability criterion. This could, however, be the subject of another paper, or a book.

6 Conclusions

Determining the frost resistance of concrete in a structure is not an easy matter. On the contrary, it has some pitfalls, of which the most serious one is obtaining a relevant number of relevant test specimens. In the ideal case, the test specimens should be 9 in number, being 100 × 100 × 400 mm prisms. They can be used to determine the relative dynamic modulus of elasticity (and, from it, RDM) as well as flexural strength, which is used to calculate the frost resistance coefficient. To this end, however, 9 core samples had to be taken from the structure with a diameter of 160–180 mm and length of at least 450 mm. The cores would also have to be without reinforcement and cracks, which, in a vast majority of structures, is next to impossible. Another (feasible) variant is using cores of a smaller diameter (such as 100 mm) to prepare test cylinders 150 mm or 200 mm long. These can then be used to determine the RDM while testing the flexural strength can be replaced by testing the tensile splitting strength.

It should be stressed that, in any case, to prepare the test specimens, core sampling and circular cutting will be required, which will inevitably entail a risk of micro cracks occurring in the test specimen concrete. This, along with the fact that the surface of the test specimens will be created by a cut with exposed aggregate, can have a negative impact on the frost resistance test. A serious problem, especially in older structures, can also be the considerable variability of concrete in its volume or the use of different concretes in different (even if structurally identical) parts of the construction—as was the case in the above experiment.

The result of a concrete frost resistance test can be used as one of the basic parameters in assessing concrete structure repair projects in terms of sustainability. It would enter the calculation of the sustainability indicator k_{SB} as a performance quantity *R*. The only condition is that the frost resistance parameter substituted into the k_{SB} calculation should be determined using the same method for all concretes.

Acknowledgements This chapter has been written as a part of project No. 19-22708S, supported by the GACR—Czech Science Foundation.

References

1. Hrabová K, Vymazal T (2019) Quantification of sustainability of concrete structures. Int Multidisc Sci GeoConference SGEM **19**/6.3
2. European Commission Energy, Strategy-Secure competitive and sustainable energy [online]. available from: https://ec.europa.eu/energy/en/topics/energystrategy
3. The 17 goals, The Global Goals for Sustainability Development [online]. available from: https://www.globalgoals.org/
4. Hrabová K, Teplý B, Hájek P (2019) Concrete, sustainability and limit states. In: IOP conference series: earth and environmental science, vol 290
5. Neville MA (2011) Properties of concrete, 5th edn. Pearson, England
6. ISO 15392 Sustainability in building construction—General principles. International Organization for Standardization, Geneva Switzerland (2008)
7. ISO 13315-1 Environmental management for concrete and concrete structures, Part 1: General Principles, International Organization for Standardization, Geneva Switzerland (2012)
8. ISO 13315-2 Environmental management for concrete and concrete structures Part 2: System boundary and inventory data. International Organization for Standardization, Geneva Switzerland (2014)
9. Fib Bulletin 71. Integrated Life Cycle Assessment of Concrete Structures, International Federation for Structural Concrete (fib), Lausanne Switzerland (2014)
10. Teplý B, Vymazal T, Rovnaníková P (2018) Introduction to an approach to performing sustainability quantification of concrete structures. In: Solid state phenomena, vol 272
11. Vořechovská D, Šomodíková M, Podroužek J, Lehký D, Teplý B (2017) Concrete structures under combined mechanical and environmental actions: modelling of durability and reliability. Comput Concr 20(1):99–110
12. Müller HS, Haist M, Vogel M (2014) Assessment of the sustainability potential of concrete and concrete structures considering their environmental impact, performance and lifetime. Constr Build Mater 67:321–337
13. Muller HS, Haist M, Moffatt JS, Vogel M (2017) Design, material properties and structural performance of sustainable concrete. Procedia Eng 171:22–32
14. Vymazal T, Teplý B, Hrabová K (2019) The assessment and prediction of concrete sustainability. In: IOP conference series: materials science and engineering, vol 583
15. ASTM C666/C666M—15 (2015) Standard test method for resistance of concrete to rapid freezing and thawing. West Conshohocken, PA, ASTM International
16. CEN/TR 15177 (2006) Testing the freeze-thaw resistance of concrete—internal structural damage. European Committee for Standardization, Brussels
17. ČSN 73 1322 (1969) Determination of frost resistance of concrete (in Czech). UHM, Prague
18. RILEM TC 176-IDC (2004) Internal damage of concrete due to frost action. RILEM Publications SARL 37/274
19. Collepardi M (2006) The new concrete, 1st edn. Grafiche Tintoretto, Treviso
20. Aïtcin P-C (1998) High-performance concrete. In: Stern B (ed). New York, E&FN Spon
21. Shang H-S, Yi T-H (2013) Freeze-thaw durability of air-entrained concrete. Sci World J 2013
22. EN 206 + A1 Concrete—Specification, performance, production and conformity. European Committee for Standardization, Brussels (2016)

23. EN 12504-4 Testing concrete—Part 4: Determination of ultrasonic pulse velocity. European Committee for Standardization, Brussels (2004)
24. ČSN 73 1372 Non-destructive testing of concrete—Testing of concrete by resonance method (in Czech). ÚNMZ, Prague (2012)
25. EN 12390-6 Testing hardened concrete—Part 6: Tensile splitting strength of test specimens. European Committee for Standardization, Brussels (2000)
26. EN 13791 Assessment of in-situ compressive strength in structures and precast concrete components. European Committee for Standardization, Brussels (2007)
27. Kocáb D, Lišovský M, Žítt P (2019) Experimental determination of free-ze-thaw resistance in self-compacting concretes. In: IOP conference series: materials science and engineering, vol 549
28. Kocáb D, Vymazal T, Kucharczyková B, Daněk P, Halamová R, Hanuš P (2018) Influence of coarse aggregate grain size on frost resistance of concrete. Key Eng Mater 776

The Application of Monte Carlo Method in the Modeling of Concrete Cover Damage in Reinforced Concrete

T. Krykowski

Abstract This paper presents the application of the Monte Carlo method to evaluate the cracking time of cover in concrete specimens subjected to accelerated corrosion testing. The distribution of probability density of some material parameters was assumed to comply with the Gaussian distribution. The process of cover cracking was analyzed for extreme values of volume strains resulting from corrosion $inf\left(\varepsilon_{ij}^{V}\right)$ and $sup\left(\varepsilon_{ij}^{V}\right)$ values determined with the Monte Carlo method.

Keywords Corrosion · Cover cracking · Monte-Carlo · Elastic-plastic-fracturing · FEM

1 Introduction

Cracking time of reinforced concrete elements caused by corrosion of the reinforcement has been widely analyzed in scientific papers. The difficulty in modeling cracks caused by corrosion of the reinforcement was connected with second and third stages of the structure degradation (the first stage refers to transport of aggressive substances, carbon dioxide and chloride ions [1, 2]). A three-stage model of degradation assumes that the initiation of electrode process [3–5] is not equivalent to the formation of the strains and stress fields in the cover. At first, filling was observed for voids in the steel-concrete interface characterized by higher porosity (the so called interface transition zone, ITZ [5]) and voids created by solubilisation of reinforcing steel during the process of reinforcement corrosion [6]. As corrosion progressed, micro-damage in the interfacial transition zone in concrete [7] occurred as the effect of increased volume strains caused by electrochemical reactions after filling pore structures, which retarded the formation of cracks in the cover.

T. Krykowski (✉)
Department of Mechanics and Bridges, Silesian University of Technology, Akademicka 5, 44-100 Gliwice, Poland
e-mail: tomasz.krykowski@polsl.pl

Z. Zembaty et al. (eds.), *Environmental Challenges in Civil Engineering*, Lecture Notes in Civil Engineering 122,
https://doi.org/10.1007/978-3-030-63879-5_6

Many models can be used to predict the time of damage in concrete cover. They include the traditional models [8], models including the time of filling pore voids [9], or models including the additional loss related to solubilisation of reinforcing steel due to electrode reaction on the rebar surface [6]. Finally, there are models that additionally include the possibility of micro-damage formation at the early stage of concrete degradation [7, 10]. Another group of publications focuses on numerical modelling of issues related to cracking of concrete covers. Some of them discuss the propagation of damage in covers [11–13] and other papers also describe the effect of many various processes (diffusion, transport of moisture, heat) [14] on propagation of that type of damage. The majority of such papers develop existing concepts of corrosion in covers by including additional physical and chemical phenomena.

The basic drawback of many models presenting aspects of cover damage is their deterministic nature and uncertainty of parameters describing corrosion. The evaluation of characteristics of material parameters for microstructure of ITZ and chemical composition of corrosion products seems to be difficult. Considering such parameters for models as random variables can solve the above difficulties. Corrosion parameters regarded as uncorrelated random variables at uniform distribution were described, inter alia, in the paper [11].

That paper presents the application of the Monte Carlo method to evaluate the cracking time of cover in concrete specimens subjected to accelerated corrosion testing. Parameters of the model were assumed to be random variables at Gaussian distribution. The probabilistic approach was used to determine boundary coordinates for the tensor of volume strains that was produced by deposition of corrosion products in the steel-concrete $inf\left(\varepsilon_{ij}^{V}\right)$ and $sup\left(\varepsilon_{ij}^{V}\right)$ interfacial transition zone.

2 Model of Reinforcement Corrosion

Initiation of reinforcement corrosion time and activation of mechanical damage in the cover are not identical. In the initial stage filling of empty pore spaces in the transition zone with higher porosity can be only observed [5]. Next along with the increasing of mechanical interactions the micro-cracks formed in the cover bordered with the transition zone between steel and concrete are being fulfilled [7]. Kinetics of such changes related to the impact of corrosion products on concrete was defined in the paper [2] in accordance with the following relationship

$$\dot{V}_{eff} = \dot{V}_{ekw} - \dot{V}_{por} - \dot{V}_{tran}, \quad \dot{V}_{ekw} = \dot{V}_R - \dot{V}_{Fe^{2+}}, \quad \dot{m}_{Fe}^{2+} = kI. \tag{1}$$

where $\dot{V}_{eff}$, $\dot{V}_{ekw}$ are respectively the effective (equivalent) rate of change in volume of corrosion products, $\dot{V}_{por}$—the rate of volume change of corrosion products penetrating into micro-cracks in the ITZ layer, $\dot{V}_{tran}$—the rate of change of volume of corrosion products transferred into deeper layers of concrete, $\dot{V}_R$, $\dot{V}_{Fe^{2+}}$—are respectively the rate of change of volume of rust (Fe ions) created in the electrochemical reaction, $\dot{m}_{Fe}^{2+}$—the velocity of the mass change of Fe ions, k—electrochemical equivalent of Fe, and I —the electric current intensity. Due to problems with determining $\dot{V}_{por}$ and $\dot{V}_{tran}$, that relationship can be expressed as the modified relationship depending on the function β which defines the intensity at which the developing corrosion affects the cover structure [10]

$$\dot{V}_{eff} = (1-\beta)\dot{V}_{ekw}, \quad \dot{V}_{ekw} = \psi I, \quad \beta\dot{V}_{ekw} = \dot{V}_{por} + \dot{V}_{tran} \tag{2}$$

$$\psi = \frac{\left(\alpha^{-1}\vartheta - 1\right)k}{\varrho_{Fe^{2+}}}, \quad \alpha = m_{Fe^{2+}} m_k^{-1}, \quad \vartheta = \varrho_{Fe^{2+}} \varrho_k^{-1} \tag{3}$$

where $\rho_{Fe^{2+}}$ is density of Fe ions mass, ρ_k—density of corrosion products, α and ϑ—adopted parameters depending on the chemical composition of corrosion products [8]. The analysis of corrosion processes of the reinforcement presented in this paper was based on the assumption that parameters describing deposition of corrosion products on the reinforcement surface were not deterministic. Parameters defined in this paper as uncertain parameters were assumed to be random variables like i.a. parameters describing chemical composition of corrosion products α, ϑ [8]. The impact of corrosion products on concrete cover changing over time was expressed using the following function β [7]

$$\beta = \begin{cases} 1 \quad for \quad t < t_0 \\ (t_{cr} - t)/(t_{cr} - t_0) \quad for \quad t_0 \le t \le t_{cr} \\ 0 \quad for \quad t > t_{cr} \end{cases} \tag{4}$$

where t_0 [15] is the time required for filling pore voids in the ITZ layer, t_{cr}—the time after which the total volume of corrosion products affects the cover structure [7, 10]. Deformations were assumed to occur only in the plane perpendicular to the axis of the rebar in accordance with the following relationship [10], [16]

$$\dot{\varepsilon}_{\alpha\beta}^{V} = \frac{\lambda(1-\beta)\left(\alpha^{-1}\vartheta - 1\right)kI}{\eta\rho_{Fe^{2+}}V_0}\delta_{\alpha\beta}, \quad \dot{\varepsilon}_{33}^{V} = 0, \quad \alpha, \beta = 1, 2, \tag{5}$$

$$\eta = 3\beta + 2(1-\beta), \quad t \le t_{cr}; \quad \eta = 2, \quad t > t_{cr} \tag{6}$$

The parameter λ in the equation is related to the loss of reinforcement corrosion products that can be washed away into the electrolyte vessel (e.g. tap water in which the sample is immersed) during the accelerated corrosion test.

The formulation of the material model was based on the strain decomposition $\boldsymbol{\varepsilon}$ into elastic $\boldsymbol{\varepsilon}^e$, plastic $\boldsymbol{\varepsilon}^p$, fracturing $\boldsymbol{\varepsilon}^f$ and the volume part related to corrosion products $\boldsymbol{\varepsilon}^V$. The rate form of constitutive equation was defined according to the well-known general formula [17]

$$\dot{\boldsymbol{\sigma}} = \boldsymbol{C} : \dot{\boldsymbol{\varepsilon}}^{e} = \boldsymbol{C} : \left(\dot{\boldsymbol{\varepsilon}} - \dot{\boldsymbol{\varepsilon}}^{p} - \dot{\boldsymbol{\varepsilon}}^{f} - \dot{\boldsymbol{\varepsilon}}^{V}\right), \tag{7}$$

where $\boldsymbol{C}$ is elastic tensor of material.

3 Determining Parameters for the Model Using the Monte Carlo Method

Material parameters for the model were randomly selected (using the Monte Carlo simulation) for the purpose of this paper. Calculations were made to evaluate the impact of corrosion products on an increase in volume strains. Material parameters were considered as dependent parameters and were to be determined in accordance with the approach described in the paper [18] (correlated random variables could be determined on the basis of uncorrelated variables) using the following relationship

$$\boldsymbol{X} = \boldsymbol{T} \cdot \boldsymbol{Y}, \quad \boldsymbol{X}^{\mathrm{T}} = \{X_1, X_2, \ldots, X_n\}, \quad \boldsymbol{Y}^{\mathrm{T}} = \{Y_1, Y_2, \ldots, Y_n\}, \tag{8}$$

where $\boldsymbol{X}$ is the matrix, whose elements are correlated random variables, $\boldsymbol{T}$ is the transformation matrix, $\boldsymbol{Y}$ is the matrix, whose elements are uncorrelated random variables. The transformation matrix $\boldsymbol{T}$ can be determined by analysing the eigen problem of the covariance matrix $\boldsymbol{C}_x$ (eigenvectors of the matrix $\boldsymbol{C}_x$ are its elements $\boldsymbol{T}$). The covariance matrix $\boldsymbol{C}_y$ for uncorrelated random variables $\boldsymbol{Y}$ and the vector of expected values for uncorrelated random variables $\boldsymbol{Y}_\mu$ can be determined from the following relationship:

$$\mathbf{C}_y = \mathbf{T}^{\mathrm{T}}\mathbf{C}_x\mathbf{T}, \tag{9}$$

$$\boldsymbol{Y}_\mu = \mathbf{T}^{\mathrm{T}}\boldsymbol{X}_\mu, \tag{10}$$

where $\boldsymbol{X}_\mu$ is the vector of expected values for correlated random variables. Elements of the covariance matrix for uncorrelated random variables were defined by the following relationship

$$C_{X_{ik}} = [cov(X_i, X_k)] = \rho_{X_{ik}}\sigma_{X_i}\sigma_{X_k}. \tag{11}$$

In the Eq. (11), $\rho_{X_{ik}}$ are coefficients of correlation of correlated random variables X_i.

The distribution of correlated random variables $\boldsymbol{X}$ can be determined by the relationship (8). In this case, the vector $\boldsymbol{Y}$ is the matrix with rows containing elements that are randomly selected uncorrelated random variables with expected values μ_{Y_i} and standard deviation σ_{Y_i} (those distributions were determined using generators of random numbers integrated in Matlab software). Value $\boldsymbol{X}$ is the matrix with rows, whose elements are randomly selected values of correlated random variables with the expected value μ_{X_i} and standard deviation σ_{X_i}.

4 Analysis of the Calculation Example

4.1 Assumption for the Model

The evaluation was made for the width of the crack in a cube (Fig. 1a) with dimensions of $80 \times 100 \times 100$ mm reinforced with a steel bar ϕ20, length L = 8 cm, connected into an electrolyzer generating 20 V DC. Voltage was kept for 17 days and measurements were recorded every minute. Results of the above tests were presented in the paper [11]. Changes in current intensity and electrical resistance of the scheme is shown in Fig. 1. FEM model of the specimen and the support scheme are presented in Fig. 2.

Calculations were performed in two stages. In the first stage, limit values of volume strain increments caused by corrosion were determined using the Monte Carlo method.

The next stage, after taking into account the above mentioned increments in volume strains, included FEM calculations to describe the propagation of crack width in the test reinforced concrete element. The interfacial transition zone between steel and concrete was included by introducing interface elements [17] to the steel-concrete interfacial transition zone. Loads on the model were considered by forcing volume strains of the rebar caused by corrosion in the process similar to thermal deformations. The combined method (Newton Raphson and Arc Length) was used for calculations assuming the following convergence conditions for displacement, out-off-balance forces and energy: $\varepsilon_{rel,disp} = \varepsilon_{rel,force} = \varepsilon_{rel,energy} = 0.01$, whereas the condition for forces in terms of maximum components $\varepsilon_{abs,force} = 0.001$ [17]. The critical time was accepted as equal $t_{cr} = 72.282$ h.

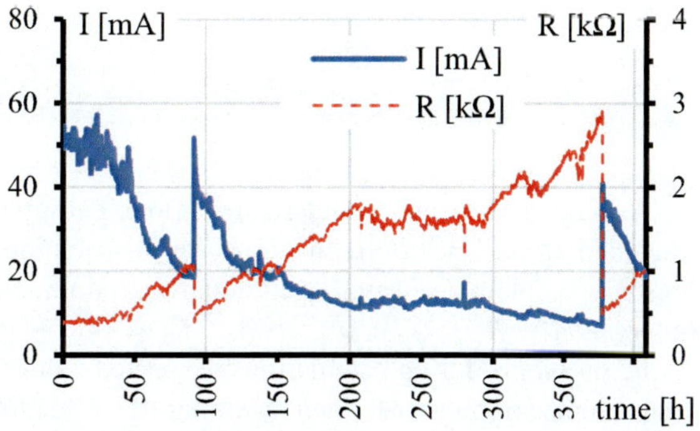

Fig. 1 Changes in current intensity and electrical resistance on the rebar surface [11]

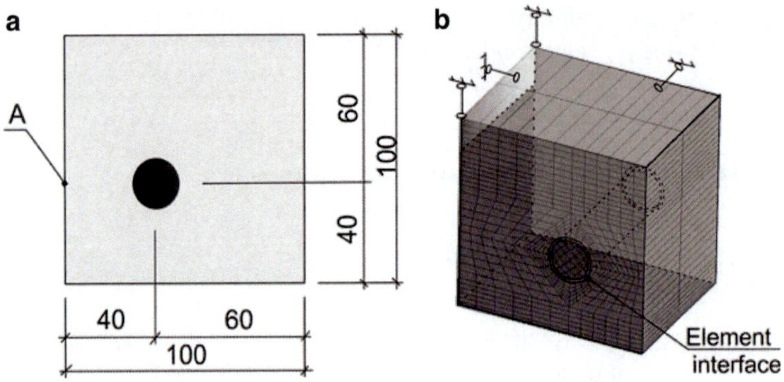

Fig. 2 Reinforced concrete specimen and FEM model: **a** dimensions of reinforced concrete specimen analyzed in the paper [11], **b** FEM model of reinforced concrete specimen with the support scheme and position of interface elements

4.2 Material Parameters

For the analyzed case, random variables were assumed to take values $X_1 = X_k$, $X_2 = X_\alpha$, $X_3 = X_\vartheta$, $X_4 = X_\varepsilon$, $X_5 = X_{ws}$. Resulting corrosion products were considered as the ideal mixture of hydroxides $Fe(OH)_2$ and $Fe(OH)_3$. Expected values μ_α and μ_ϑ for parameters α and ϑ were adopted as average values on the basis of the paper [8]: for iron (III) hydroxide, $\alpha_{Fe(OH)_3} = 0.523$, $\vartheta_{Fe(OH)_3} = 2.09$, and iron (II) hydroxide, $\alpha_{Fe(OH)_2} = 0.622$, $\vartheta_{Fe(OH)_2} = 2.24$. Expected values for other random variables were arbitrarily adopted on the basis of data available in the literature. As parameters of the model could not be experimentally evaluated, values of standard deviation determined for random variables at the uniform distribution $X \in \langle X^-, X^+ \rangle$, average values $X_0 = 0.5\big(X^+ + X^-\big)$ and the interval $\Delta X = X^+ - X^-$, were used for calculations, assuming the deviation from the average by $\eta = 5\%$ and $\eta = 10\%$ in accordance with the following relationship

$$X^- = X_0 - 0.5\Delta X, \quad X^+ = X_0 + 0.5\Delta X, \quad \Delta X = \eta X_0 \tag{12}$$

$$\mu_{X_i}^{unif} = \frac{X_{max} + X_{min}}{2}, \quad \sigma_{X_i}^{unif} = \sqrt{\frac{(X_{max} - X_{min})^2}{12}}, \tag{13}$$

Average values and standard deviations determined as specified above, were regarded as parameters of the Gaussian distribution, for whom material parameters were randomly selected. Additionally, random selection of values was limited to intervals R^-, R^+, $R^- = \mu_{X_i} - 2\sigma_{X_i}$, $R^+ = \mu_{X_i} + 2\sigma_{X_i}$ (95.4% of observations) to avoid problems during calculations. Expected values μ_{X_i}, standard deviations, σ_{X_i} and assumed minimum R^- and maximum R^+ values of the distribution are compared in tab. 1 and 2 for parameters obtained for the uniform distribution at 5% deviation

Table 1 Comparison of average values, standard deviations and extreme values of distribution for parameters calculated according to scheme A

Corrosion parameters	Expected value	Stand. deviation	Inf.	Sup.
	μ_X	σ_X	R^-	R^+
Parameter, α (1)	0.573	0.0083	0.556	0.589
Parameter, ϑ (1)	2.165	0.0313	2.103	2.227
Porosity of transitional zone, ε (1)	0.550	0.0079	0.534	0.566
Width of transitional zone [15], $w_{wp} \cdot 10^3$ (cm)	7.500	0.1083	7.283	7.717
Electrochemical equivalent of Fe, $k \cdot 10^3$ (g/μ Arok)	6.308	0.0911	6.126	6.490

Table 2 Comparison of average values, standard deviations and extreme values of distribution for parameters calculated according to scheme B

Corrosion parameters	Expected value	Stand. deviation	Inf.	Sup.
	μ_X	σ_X	R^-	R^+
Parameter, α (1)	0.573	0.0165	0.539	0.606
Parameter, ϑ(1)	2.165	0.0625	2.040	2.290
Porosity of transitional zone, ε (1)	0.550	0.0159	0.518	0.582
Width of transitional zone [15], $w_{wp} \cdot 10^3$ (cm)	7.500	0.2165	7.066	7.933
Electrochemical equivalent of Fe, $k \cdot 10^3$ (g/μ Arok)	6.308	0.1821	5.944	6.672

from the average value (case A) and 10% deviation from the average value (case B) (Tables 1 and 2).

Parameters used for calculating the increment in volume strains were assumed to be correlated. For the purpose of calculations, the matrix of random variable correlation coefficients $\rho_{\alpha\alpha} = 1$, $\rho_{23} = \rho_{32} = 0.5$, $\rho_{45} = \rho_{54} = 0.5$ (other values are zero) has been arbitrarily adopted. The Monte Carlo method for N = 500 randomly selected numbers was used in calculations. As a result of the calculations, the processes of changes in the equivalent volume and the effective volume of corrosion products were obtained, respectively for cases A and B considered in the work Figs. 3 and 4, and the sample load program for volume strain increments, Fig. 5. Total value of volume strains calculated in accordance with Eq. (5) was calculated in 10 consecutive time increments for calculation reasons. The concrete sample loading program with the mentioned increases was presented for the lower $inf\left(\Delta\varepsilon^V\right)$ and upper $sup\left(\Delta\varepsilon^V\right)$ curves, respectively, limiting the allowable increments in the volume of corrosion products.

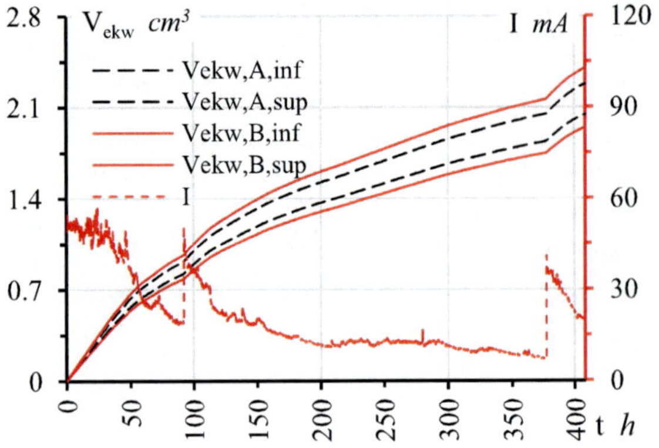

Fig. 3 Changes in equivalent volume of corrosion products determined with the Monte Carlo method (case A and B)

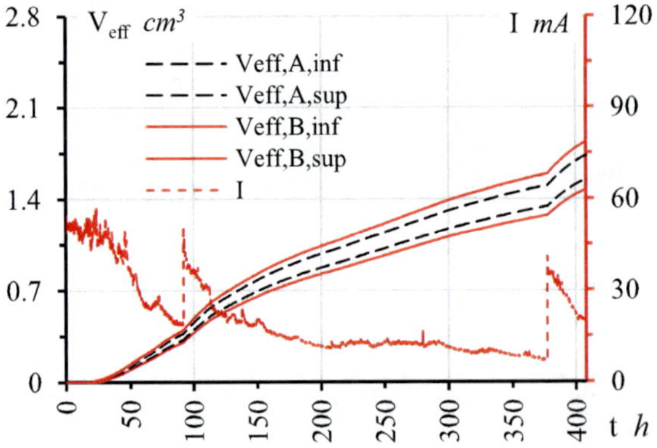

Fig. 4 Changes in effective volume of corrosion products determined with the Monte Carlo method (case A and B)

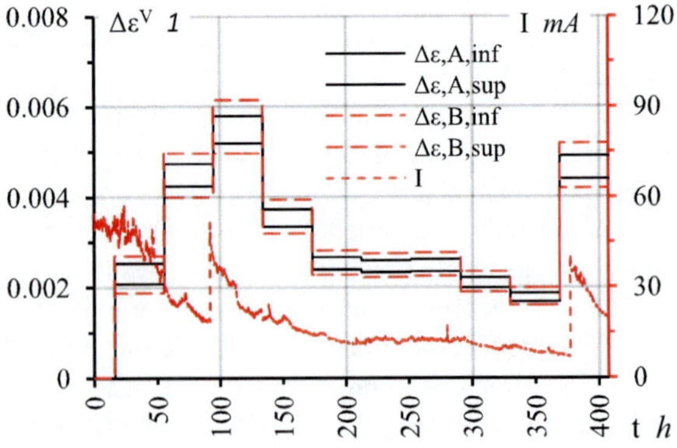

Fig. 5 Limit increments of effective strains of corrosion products determined using the Monte Carlo method $inf(\Delta\varepsilon^V)$ i $sup(\Delta\varepsilon^V)$ (case A and B)

Table 3 Parameters of concrete for FEM calculations [11]

Modulus of elasticity, E (GPa)	38.28
Poisson's ratio, υ (1)	0.2
Average tensile strength, f_{ctm} (MPa)	3.99
Average compressive strength, f_{cm} (MPa)	56.4
Fracture energy, G_F (kN/m)	0.151
Maximum diameter of aggregate particles d_{max} (mm)	8
Active shear factor (1)	20

Table 4 Material parameters for interface elements (steel-concrete interface transition zone)

Normal stiffness, k_{nn} (MN/m^3)	$2 \cdot 10^8$
Tangential stiffness, k_{tt} (MN/m^3)	$2 \cdot 10^8$
Cohesion, c (MPa)	1
Friction coefficient, ϕ (1)	0.1
Tension strength, f_t (MPa)	0.3

Table 5 Parameters of steel for FEM calculations [11]

Modulus of elasticity, E (GPa)	200
Poisson's ratio, υ (1)	0.3
Yield strength of steel, f_y (N/mm^2)	235

Values showing the width of cracks in the reinforced concrete specimen were determined in the next stage. Calculations were made for the adopted Monterey-Willam and Rankine-Fracturing material model for concrete and perfectly elastic -plastic model by Huber-Misses-Hencky for steel. Material parameters for concrete and steel are shown in Tables 3 and 4. Parameters necessary for defining the interfacial transition zone are shown in Table 5.

As a result of the calculations, the courses of changes in the crack opening width in the sample as a function of time were determined for the lower $inf(\Delta\varepsilon^V)$ and upper $sup(\Delta\varepsilon^V)$ limits of the volume increase of corrosion products. The results of the crack opening width were determined at the Gauss points of the finite element (at the Gauss point located closest to the point measuring the crack opening A). The calculation results obtained for both considered load cases and the graphic image visualizing the way the sample is scratched assuming the visibility of the scratches meeting the criterion $w_{min} \geq 0.1$ mm is presented in Fig. 6.

5 Summary and Final Conclusions

The proposed calculation approach is useful in estimating intervals, within which damage to reinforced concrete elements in structures are observed under the corrosion

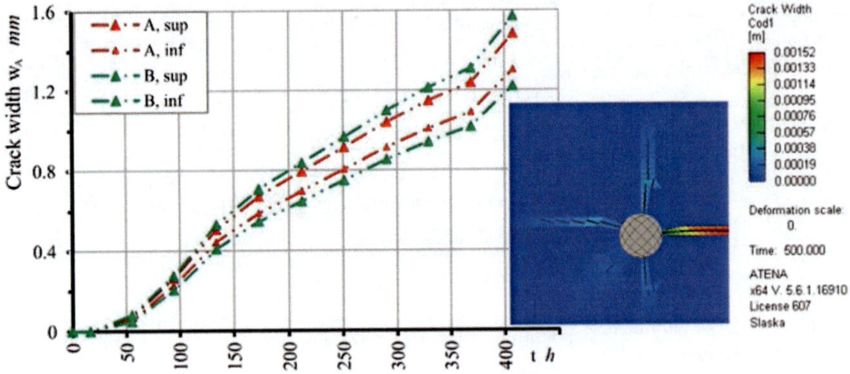

Fig. 6 Limits of crack widths in the reinforced concrete test element for limiting increases in corrosive volume strains determined using the Monte Carlo $inf\left(\Delta\varepsilon^V\right)$ and $sup\left(\Delta\varepsilon^V\right)$ methods, together with the map of scratches, diagrams A and B

of reinforcement. The suggested scheme of procedure should be regarded as the indicative approach. The function of distribution of probability density, which should be experimentally determined, is crucial for such calculations. This paper describes the approach to determine the interval as the lower $\inf(w)$ and upper $sup(w)$ bounds, at which cracks of specified width can appear over the function of time, Fig. 6. Applying this approach in calculations can also be useful in evaluating the time interval, at which cracks having the width that is critical considering the further use of the structure.

References

1. Koniorczyk M, Gawin D (2006) Numerical modeling of coupled heat, moisture and salt transport in porous materials. Comput Assist Mech Eng Sci 13(4):565–574
2. Suwito C, Xi Y (2008) The effect of chloride-induced steel corrosion on service life of reinforced concrete structures. Struct Infrastruct Eng Maintenance Manage Life-Cycle Des Perform 4(3):177–192
3. Dao LTN, Dao VTN, Kim S-H, Ann KY (2010) Modeling steel corrosion in concrete structures—part 2: a unified adaptive finite element model for simulation of steel corrosion. Int J Electrochem Sci 5:314–326
4. Bazant ZP (1979) Physical model for steel corrosion in concrete sea structures—theory. J Struct Div ASCE 105:1137–1153
5. Bentur A, Alexander MG (2000) A review of the work of the RILEM TC 159-ETC: Engineering of the interfacial transition zone in cementitious composites. Mater Struct 33:82–87
6. Martín-Pérez B (1999) Service life modeling of R.C. highway structures exposed to chlorides. Ph.D. Thesis
7. Michel A, Pease BJ, Peterová A, Geiker MR, Stang H, Thybo AEA (2014) Penetration of corrosion products and corrosion-induced cracking in reinforced cementitious materials: experimental investigations and numerical simulations. Cement Concr Compos 47:75–86

8. Pantazopoulou SJ, Papoulia KD (2001) Modeling cover-cracking due to reinforcement corrosion in RC structures. J Eng Mech 127:342–351
9. Liu Y, Weyers RE (1998) Modeling the time-to-corrosion cracking in chloride contaminated reinforced concrete structures. ACI Materials Journal 95:675–680
10. Wieczorek B, Krykowski T (2017) Application of damage mechanics rules to evaluate the growth of corrosive deformations in transition layer. Ochrona przed Korozją 1:3–6
11. Krykowski T, Jaśniok T, Recha F (2019) Modelowanie czasu uszkodzenia otuliny betonowej z uwzględnieniem niepewności parametrów modelu. Ochrona przed Korozją 62(3):69–73
12. Andrade C, Alonso C, Molina FJ (1993) Cover cracking as a function of bar corrosion: Part I-experimental test. Mater Struct 26:453–464
13. Molina FJ, Alonso C, Andrade C (1993) Cover cracking as a function of rebar corrosion: Part 2-numerical model. Mater Struct 26:532–548
14. Ožbolt J, Oršanić F, Balabanić G (2017) Modelling processes related to corrosion of reinforcement in concrete: coupled 3D finite element model. Struct Infrastruct Eng Maintenance Manage Life-Cycle Des Perform 13:135–146
15. Liu Y (1996) Modeling the time-to-corrosion cracking of the cover concrete in chloride contaminated reinforced concrete structures, Virginia Polytechnic Institute and State University, Blacksburg, Virginia
16. Krykowski T (2020) The application of interval/affine algebra to determine the time of concrete cover damage in reinforced concrete due to corrosion. Comput Assist Methods Eng Sci (CAMES) 27(4):265–284
17. Červenka V, Jendele L, Červenka J (2016) Atena Program Documentation, part 1, theory. Červenka Consulting, Prague
18. Nowak AS, Collins KR (2012) Reliability of structures, 2nd edn. CRC Press

Discovery of the Oldest Construction Phases of the Convent of the Order of St. Dominic in Sandomierz—Research and Technical Conditions of Their Conservation and Restoration

A. Kadłuczka and K. Stala

Abstract Following the ages-old trade route from Western Europe through Krakow and Lviv towards the Kievan Rus', when nearing Sandomierz—one of the Piast dynasty's *sedes regni principalis*—the traveller's attention will first be drawn towards the brick, compact massing of the monastery of the Order of Saint Dominic and, in the background, the monumental massing of a church, dominating in the skyline of the Hill of St. Jacob. The convent of the Order of St. Dominic in Sandomierz is indeed a striking documentation of the perfection of architectural technique and proof of cultural links and contacts with the northern and western circles of European construction and artistic schools. Adam Miłobędzki, a well-known scholar of Polish architecture, stated outright: "*The church in Sandomierz belongs to some of the best cases of early Dominican architecture in Europe*", describing its remarkable ceramic decoration as "sophisticated" and bearing the marks of northern Italian buildings. Similarly, Jan Zachwatowicz, in his synthetic publication entitled Architektura Polska, published in 1966 but written much earlier, characterised the rich ceramic decoration of the Dominican complex as being of the Lombard type, simultaneously pointing towards the archetypical alternation of architectural and decorative forms. Zygmunt Świechowski noted the stylistic and technical diversity of the presbytery and the nave, parts that some scholars consider uniform in terms of their spatial proposal. The building technique of the first section was observed to operate using simpler means of expression, which can be seen in its bonding, the manner of the laying of the arcaded frieze, with diagonally placed bricks, while in the other—the nave—we can see a more sophisticated and technologically improved ceramics. In the literature we can find a well-entrenched view—based on surviving plans of the convent from 1864, drafted just before its partial dismantlement—which assumes that the initially planned complex did not differ substantially from the shape that survived up to the middle of the nineteenth century. Meanwhile, archaeological research conducted in the years 2016–2018 brought a series of surprising discoveries, casting a completely new light on the beginnings of the Dominican monastery, confirming

---

A. Kadłuczka (✉) · K. Stala
Faculty of Architecture, Institute of History of Architecture and Monument Preservation, Cracow University of Technology, Kraków, Poland
e-mail: a-1@pk.edu.pl

Z. Zembaty et al. (eds.), *Environmental Challenges in Civil Engineering*, Lecture Notes in Civil Engineering 122,
https://doi.org/10.1007/978-3-030-63879-5_7

the stylistic and technical diversity of its individual parts, while also showing the mission of the Order of St. Dominic in Poland at the time, which combined religious goals with political ones, in a completely new light, revealing the crucial role of the monastery in Sandomierz. It was meant to be a coordination centre for the Order's initially broad operations in the territory of the Kievan Rus', which ceased because of the destructive Tartar invasions in 1241 and 1259. Traces of the initial layout were discovered in the remains of the southern wing, which was demolished in the nineteenth century, and which had initially been located further southwards, enclosing a larger, square-shaped cloister garth, as well as in the bonding of the so-called Romanesque cellar with a single column, located in a corner of the eastern and southern wing, which remained after the reduction of the functional and spatial programme of the two-segment chapterhouse. The area where the southern wing and the south-western corner of the monastic complex were located is a strategic site, both from an academic point of view and in the aspect of the spatial recomposition and conservational renovation of this exceptional site of Polish Medieval architectural heritage. The archaeological and architectural research that has thus far been performed, despite its limited extent, enables the delineation of strict protection zones and areas where further research should develop, in addition to initiating design work on restoring the historical spatial composition based on contemporary conservation doctrine, using innovative technical solutions. The chief premise of these solutions is to obtain a synergic effect as a result of the harmonious combination of existing sections and new architectural forms, along with a non-invasive technical intervention aimed at the necessary securing and preservation of heritage substance. In light of the findings that have proven crucial to the history of the Convent and medieval Polish architecture, the definition of directions and conditions for building a model of the restitution of its southern wing—which is to be programmed so as to minimise or even eliminate construction-related interference in surviving authentic historical substance—becomes paramount. Based on contemporary architectural heritage theory and an analysis of the structure's primary research nodes, the proposed technical, technological, functional and spatial solutions were based on the doctrinal priority of the historical layout's landscape recomposition and making the most precious relics that document the beginnings of the Convent accessible to the public.

Keywords Architectural heritage · Research and protection · Revalorisation

1 Introduction

Within the cycle of medieval Poland's civilisational, cultural and technological transformation, we are dealing with two key turning points of fundamental significance to academic research on Polish civil engineering and architecture, which also condition the pursuit of optimal methods of conserving and restoring historical and heritage sites that have survived in varying degrees of preservation and that are typically characterised by a complicated archaeological and architectural stratigraphy.

These two turning points are the replacement of wood—a traditional material that had been widely used in the Middle Ages—with stone, a new material that required familiarity with new construction techniques and technologies, as well as the introduction of ceramic bricks in the twelfth and thirteenth century. Its production required access to deposits of and the skill to fire clay formed into cuboid blocks, which gained much greater structural resistance and provided greater freedom in shaping architectural form as a result of being processed in high temperatures.

2 History of the Convent

Ireneusz Płuska reports that clay was used as early as in the Neolithic period to build shelters and earthen dwellings, and in the period prior to and during the early reign of the Piast dynasty it was used as the outer cladding of the timber structures of defensive gods so as to protect them from fire, but "*the oldest instance of brick construction was discovered on the grounds of the collegiate church in Tum near Łęczyca, supposedly dated to the twelfth century*" [1]. However, there can be no doubt that it is the convent of the Order of St. Dominic in Sandomierz, with the church of St. Jacob, that is "*the greatest jewel of early medieval brick architecture… built by bricklayers who had arrived directly from Lombardy*" [1], a remarkable document of technical and architectural perfection and proof of the existence of ties and cultural contacts with northern and western circles of European building and artistic schools. The use of brick blocks, particularly in the basal parts of the lowest level in the buildings of the Sandomierz convent, blocks with a doubled dimension of a standard brick, i.e. 14 × 20 × 28, often constitute the first layer of a brick wall, laid on a stone foundation. The unique architectural decoration of the Dominican church and the oldest, well-preserved eastern wing is based on serially produced cornice blocks, arches and splayed windows, surround frame lenses and friezes placed underneath eaves and on gables. The cornice blocks that were uncovered have a profile with a clearly ancient provenance, based on the composition of a convex profile (*trochilus*) and a torus, incorporated into the module of a brick block: 20 × 27–28 cm in dimensions and a thickness of 13–14 cm (Fig. 1).

The architectural value of the heritage site, described by Jan Zachwatowicz in the monumental version of Dzieje Sztuki Polskiej, in its third part—Architektura—confirms that the structure we are dealing with in Sandomierz is a model work, one that is innovative in every way and artistically perfect. He wrote: "*The church of St. Jacob in Sandomierz, along with its adjacent monastery, remains the most beautiful example of Dominican architecture in Poland*" [2], one that was built by a "*highly qualified*" team that left mason marks and inscriptions with the following names: JO[an]ES, NICOL[aus] and DANIHEL, and was probably "*of Lombardian origin, as the del Crocefisso church, a part of the San Stefano complex in Bologna, appears*

Fig. 1 Contemporary view of the monastery from the east (photo by authors)

to be the closest analogy, with related portals including that of the basilica of St. Francis in Assisi and Sagra di San Michele" [2] (Fig. 2).

Zygmunt Świechowski highlighted that "*the dissemination of this type of engineering, which had been new in Poland*", namely the ability to produce brick, apply ornament to it and erect buildings from it, was not the sole contribution of the Order of St. Dominic, which had its own cohorts of murators, but was also partly done by the Franciscans, as excellently evidenced by their Krakow monastery with an original church built on the plan of a Greek cross immediately after 1236, which bears close stylistic analogies to the eastern facade of the Sandomierz Dominican building. Świechowski (similarly to Zachwatowicz) expressed no doubt that "*Thanks to mendicant orders, forms derived from the traditional territory of brick construction – Lombardy – made their way directly to Poland. This statement is important as studies previously tied early Polish brick buildings primarily with influence from northern Germany. The reality of the matter had been much more complex.*" [3]

The European architectural standard of the Dominican Order's convent in Sanodmierz was also noted by Andrzej Buko, who argued that "*The ornamentation of the church of St. Jacob is a phenomenon that can be considered unique in Central and Eastern Europe, constituting an example of the use of an unprecedented wealth of northern Italian ornamental art*" [4].

Unfortunately, this wonderful Polish architectural heritage site greatly suffered as a result of the socio-political condition of the land of Sandomierz in the period between the seventeenth and the nineteenth century. First as a result of a war with

Fig. 2 Contemporary view of the remains of the monastery close from the south-east (photo by authors)

Sweden, then because of the loss of independence and the subsequent national uprisings and religious-political repression, the monastery and its buildings became a ruin, leading to the dismantlement of the southern wing and of a portion of the western one (Fig. 3).

3 Recent Monastery Research

The latest archaeological studies, conducted in the years 2016–2018 by Bugaj [5] and Gołembnik [6], were based on state-of-the-art research methodology, utilising "*revolutionary breakthroughs in measurement technology and digital image recording*" [6], which, in terms of programmatic assumptions, were intended to be "*an attempt at utilising available technical means to develop a documentation system enabling a secondary, virtual collation of previously removed stratigraphy along with defining the relationships between layers and any discovered structures*" [6]. Despite sceptical authorial self-assessment of this methodology's results, "*the attempt to virtually recreate the original stratigraphic layout was only partially successful*" [6] as the researchers encountered obstacles that had been difficult to foresee. What is interesting, the most significant obstacle was the imperfection of

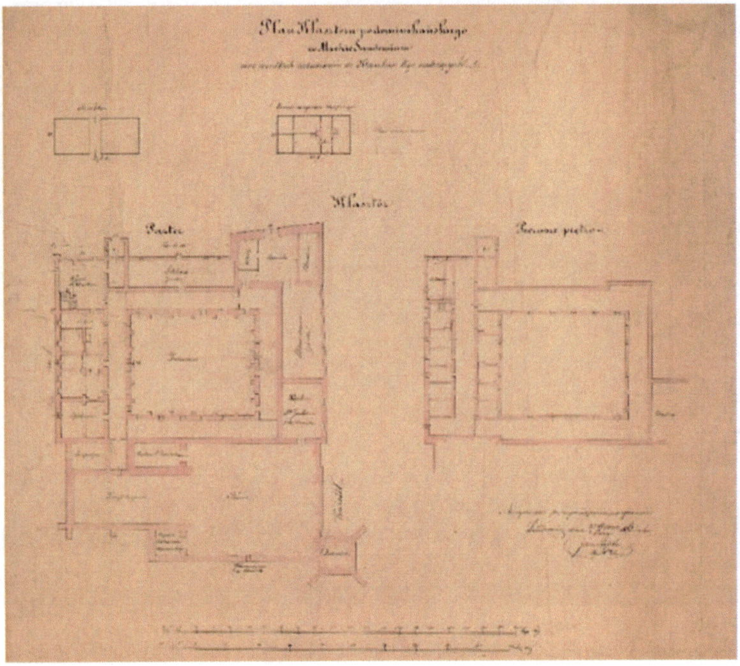

Fig. 3 Plan of the monastery layout from 1864 before demolition of the southern wing (State Archives in Radom)

the physical exploration of archaeological stratigraphy in the context of technologically advanced measurement devices, which resulted in the virtual image presenting individual planes of distortions and parametric gaps that prevented the full application of the precision afforded by the measurement instruments. Nevertheless, even in the case of faults or gaps in the database created in such a manner, substantively valuable artefacts, layers and key nodes were documented, which can and should be interpreted in an interdisciplinary manner.

The most important results of the studies performed in the years 2016–2018, which are of essential significance in the new discussion about the oldest phases of the convent's construction, include the newly discovered remains of the architectural articulation of a portal located on the internal side of the southern wall of the single-column cellar, which Andrzej Gołembnik associates with "*the abandoned intention to extend the eastern wing*", rejecting the "*functioning of the entrance, even if for a short time*" without sufficient justification [6]. The second important field of study was the area encompassing the point of contact between the southern wing, which was dismantled in the nineteenth century, and the western wall of the single-column cellar, where the relic of a node of the internal south-eastern corner of the cloister garth, initially planned *ad quadratum*, was found.

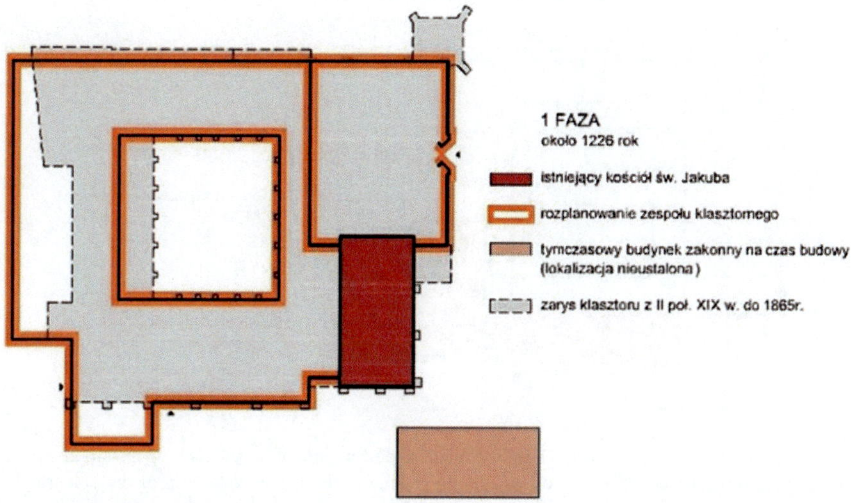

Fig. 4 Hypothetical reconstruction of the original plan of the monastery complex in Sandomierz by A. Kadłuczka and K. Stala (drawing by Archecon Studio)

Another important discovery was the unearthing of the hearth of a hypocaust-type furnace, adjacent to the external surface of the southern wing wall's foundation, along with traces of a central heating installation [6].

The relics discovered in the years 2016–2018, which belong to oldest phases of the convent's construction, when confronted with various sources and historical knowledge of the beginnings of the Dominican Order's operations in Poland, can be seen as a confirmation of its significant mission, which was not only religious, but also political in nature (Fig. 4).

Julia Tazbirowa explicitly writes that "The Dominicans had the strong support of Leszek the White and were intended to play a specific political role. Immediately after arriving in Krakow, before the consecration of the church of the Holy Trinity, Jacek Odrowąż headed to Ruthenia, where he could have played the role of a political emissary rather than a missionary. … Dominican convents should be seen as a sort of Leszek's diplomatic outposts, as he abandoned neither the intent to subjugate Pomerania, nor the spread of his influence across Ruthenia." [7] Having analysed the activity of Iwo and Jacek Odrowąż, one can assume the existence of a key role of the Order of St. Dominic's Sandomierz-based outpost in exerting influence eastwards: to the Kievan Rus', which required establishing a central command and coordination centre and the education of a missionary cadre capable of sacrifice and instilled with courage based on deep faith. The newly-established monastery in Sandomierz appeared to be a suitable place for this function as it was located on the eastern borders of Lesser Poland and along a strategic trail that led from Krakow, through Sandomierz, to Kiev. According to Jan Andrzej Spież OP [8], although the Ruthenian mission would have to be active in areas that had already been Christianised by the Eastern Church, the task of the Dominican missionaries was to convince Ruthenian

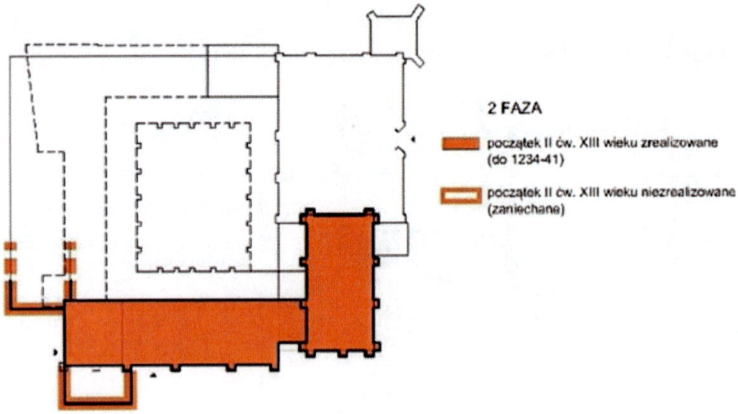

Fig. 5 Result of Tartar raids in 1241 and 1259: the abandonment of wide-reaching missionary plans, putting the ongoing construction of the southern wing on hold and reduction of the construction programme, which led to its relocation northwards and the reduction of the cloister garth by A. Kadłuczka and K. Stala (drawing by Archecon Studio)

princes and clergy to some form of union with the Holy See in Rome, as well as providing patronage over the Latin-rite Christian community which was present in Ruthenia in significant numbers. The unforeseen events that soon transpired in Kiev foiled the efforts to Christianise the Eastern Church and convince Ruthenian princes and bishops to enter a union with the Holy See, and the recurring, devastating Tartar raids: the first in 1241 and then in 1259, must have resulted in the abandonment of wide-reaching missionary plans, putting the ongoing construction of the southern wing on hold and contributing to the alteration and reduction of the construction programme, which led to its relocation northwards and the reduction of the previously planned cloister garth (Fig. 5).

Andrzej Gołembnik argues that "As a consequence of the dramatic events associated with Tartar raids, the monastery's construction cycle was put on hold twice. The repeated destruction and complete depopulation of the city over the course of merely a quarter-century, along with a new political situation that generated fear and uncertainty, had most probably had an effect on the plans of the monastery's builders" [6].

The 2016–2018 archaeological studies explicitly confirmed that the site of the monastery's southern wing is key to understanding the history of its construction and should be subjected to further study, which should also lead to exposing archaeological and architectural relics, enabling not only securing and conserving them, but also their public exhibition [9].

4 Conservation and Technical Issues

The research process, which requires both developing technological principles of securing and protecting these relics, will also necessitate providing appropriate technical conditions in the form of a cover structure that will shelter them from atmospheric conditions and will allow for regulating climate parameters inside the structure. Structures of this type appear at archaeological sites intended for so-called *musealisation* [10] and take on the form of deliberately designed covering constructions [11] in the form of a *building* or *structure* that is architectural in character. Such solutions are well-known, particularly in European ancient culture sites, where the archaeological preserve is a popular form of the *musealisation* of an archaeological site. The preserve is a deliberately arranged protective space, which forms either an enclosed, developed exhibition, or an open-air exhibition [11]. Among the many innovative examples of enclosed preserves that show signs of being outstanding architectural projects—worthy of analysis in light of the case study at hand: the Sandomierz convent of the Order of St. Dominic—we can list the large, glazed structure suspended over the remains of the Great Baths linked with the modern Römer Museum, located at the Xanten Archaeological Park near Cologne, designed by German architects Gatermann + Schossig (in cooperation with IPP Polonyi & Partner) [12], or the Neo-modernist pavilion of the Gallo-Roman Vesunna Museum in *Perigueux (Department of Dordogne), designed by Jean Nouvel, which exhibits the remains of a Roman residence that features mosaics and a rich collection of priceless artefacts. The musealisation of the archaeological site and its protection from atmospheric conditions was performed by covering it with a roof supported by 14 narrow, fourteen-metres-tall columns that rest on micropiles, which minimally interfere with the site's stratigraphy, and a glass curtain wall* [13].

Of note is the fact that a mixed ventilation system was applied here: the lower part of the glass curtain wall has a gap along its perimeter, allowing for natural gravitational ventilation of the preserve's internal space, but the building is also fitted out with a HVAC installation, which enables the fine-tuning of the environmental parameters of the space in which the relics are exhibited (Figs. 6 and 7).

Similar principles were followed in recent years in the design and construction of a branch of the Archaeological Museum—the *Kaiserpfalz franconofurd* [14] exhibition, which is a part of a larger project of the restoration of the historical centre of Frankfurt am Main, or rather the construction of a *New Old Town* there [15]. The new *Stadthaus* building, along with the *Kaiserpfalz franconofurd* archaeological exhibition, is an interesting case of contemporary architecture that is excellently incorporated into its historical context and is also a conveyor of symbolic content through a combination of the innovatively arranged relic exhibition of the imperial residence of Charlemagne with formal reception rooms of the City Hall (*Stadthaus*). The modern space of the enclosed preserve, which ensures controlled, custom climate conditions, established at the site of the former open-air *Archäologische Garten*, presents the key fragments of a residence from the two-millennia-long history of the city. The form and content of the new exhibition were determined by the results

Fig. 6 Monastery development concept with reconstruction of the southern wing and a multi-functional building at the front (modelling by Archecon Studio)

Fig. 7 Concept of architectural restitution of the southern wing demolished in the late nineteenth century (drawing by Archecon Studio)

of archaeological studies which were performed in the years 2012–2014 by local archaeological services [16], which expanded the scope of the presentation beyond that of the former *Archäologische Garten*, linking it with a new, separate exhibition organised at an exhibition salon at Bendergasse.

Insofar as the examples of enclosed preserves presented above allow the full control of interior climate parameters [11], they are a form of "*archaeological cover… understood as a type of architectural structure that provides on-site protection to relics discovered through archaeological studies… it is intended to minimise the impact of external environmental conditions that negatively affect historical substance and change because of the open character of the exhibition*" [11]. Of note is the fact that experience has shown that archaeological covers can also have an adverse effect on the relics they are meant to protect, as—depending on specific solutions and materials—"*external climate conditions can change underneath the cover, escalating temperature, humidity or exposure to sunlight and UV radiation*" [11].

Thus, this type of archaeological cover appears to be suited merely as a temporary measure in the case of the Dominican monastery in Sandomierz, to be applied for the duration of extended archaeological studies, with the rational end goal being to design an enclosed protective structure that could allow for the regulation of the indoor climate and the exhibition of the archaeological and architectural relics on the lowest level and on the ground floor, to be successively explored, conserved and exhibited, with research and technical facilities on a level above the ground floor and potentially its attic. This structure, featuring the characteristics of an architectural landmark—at the site of the non-existing southern wing and in a portion of the western one, should be their spatial restitution, restoring the historically shaped composition of the convent, and should remain in harmony with the brick walls in terms of colour, but at the same time operate using contemporary architectural form and modern detail.

Here we touch upon an important theoretical and doctrinal matter, but the precepts outlined in the Venice Charter of 1964 can be a still-relevant interpretation in this case. The Venice Charter's Article 9 reads as follows: "*The process of restoration is a highly specialized operation. Its aim is to preserve and reveal the aesthetic and historic value of the monument and is based on respect for original material and authentic documents. It must stop at the point where conjecture begins, and in this case moreover any extra work which is indispensable must be distinct from the architectural composition and must bear a contemporary stamp. The restoration in any case must be preceded and followed by an archaeological and historical study of the monument.*" [17]

Apart from theoretical, doctrinal, research, exhibition and educational considerations, construction-related and technical matters will prove to be of particular essence and will require individual, innovative solutions. However, due to the historical, academic and cultural significance of the structure in question, these solutions cannot be a technical experiment, but must instead be based on previous experiences and current knowledge while taking the individual characteristics of the archaeological site into consideration. "*The state of the preservation of the relics, the type of*

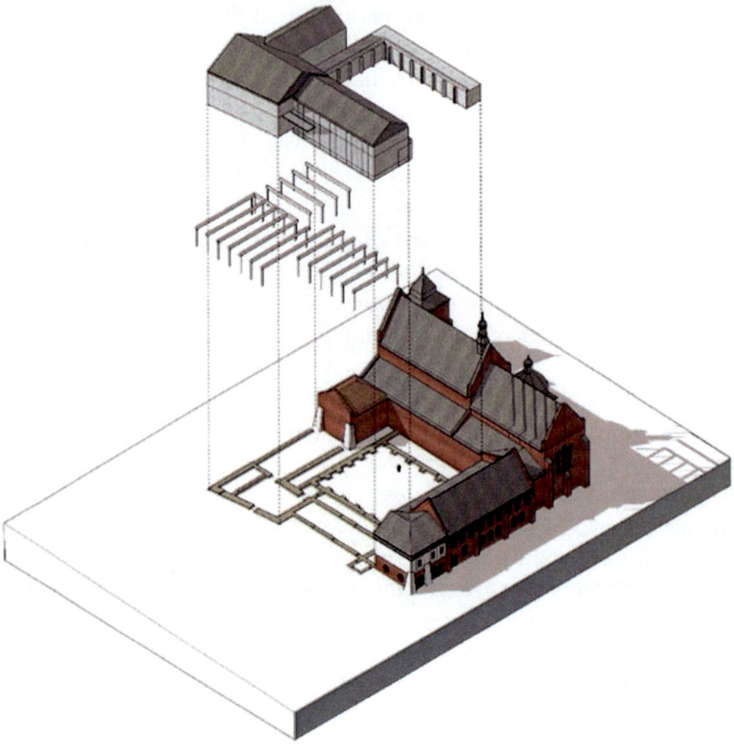

Fig. 8 Bypass-type steel structure not interfere with the relics of past buildings, and enable conducting studies after the construction of the newly-designed structure (modelling by Archecon Studio)

object or objects to be protected, their material and scale, should all be subjected to a detailed analysis. All data associated with the surrounding climate conditions are essential. Based on collected information, one can make decisions concerning the use of materials that can prove optimal both in the case of load-bearing structures, as well as the roofing itself, its anchoring, shape, scale, span or the number of necessary exposure points, their transparency, sunlight protection, cross-ventilation system, drainage, etc. The new structure must correspond with its surroundings and should substantively reference the specificity of the heritage it protects" [11] (Fig. 8).

5 Conclusions

Keeping the above in mind, two design alternatives are considered. Both assume their starting point to be giving the protective structure the form of an enclosed building that is the spatial restitution of the southern wing and of a portion of the western one, within the outline of the walls known from the plan of the monastery layout from 1864. The first alternative, in accordance with the recommendations of the Consultation and Advisory Team, assumes the construction of the footing walls of this building using traditional techniques and construction materials, to be recreated on the basis of simultaneous archaeological studies, and supported by surviving, appropriately reinforced foundations. Where no foundations are found to have survived, the footing is to be recreated in the negative of historical trenches, similarly as it was done in the case of the *Stadthaus* in Frankfurt am Main. There is a wealth of literature on this subject, in addition to numerous examples of effective solutions that abide by conservation principles [18–20].

Depending on the findings of initial archaeological studies and on the condition of determining the existing footing to be unsuitable for use (there exists a possibility of uncovering the remains of the oldest, original monastery, that perhaps was never completed), an "emergency" alternative was formulated, based on a bypass-type structure, which would not interfere with the relics of past buildings, but that would enable conducting studies after the construction of the shell state of the newly-designed structure. This is possible by designing the new structures as being "suspended" above the research area, thus making the project schedule independent of the research schedule, financial means and the various methods of investigating this area, which the design intends to constitute an archaeological and architectural preserve and a museum of both the history of the Order of St. Dominic in Poland and medieval brick architecture, featuring the necessary academic, educational and technical facilities.

The basis for the restored buildings will be a steel grid composed of transverse and longitudinal steel beams, with point supports resting on the tops of micropiles, inserted along the length of the external walls of the cellar level, approximately every 300 cm (as in the case of the Vesunna Museum in Perigeaux). The support brackets of the beams comprising the grid will be located so as to maintain a distance of around 10–15 cm from the exposed top of the walls, which will enable the application of a mixed ventilation system in the enclosed space of the preserve. The grid will be extendable as to prepare for the possible extension of the archaeological dig and the associated enlargement of the exhibition area.

Regardless of the selected design alternative, carrying out the restitution of the buildings of the former southern wing and a section of the western one will allow for not only restoring an essential element of the cultural landscape, but also protecting of the oldest relics of the monastery buildings from adverse atmospheric conditions, providing the possibility to conduct archaeological research and to gradually organise a museum exhibition so as to make its historical, academic and educational assets available to the public.

The analysis of archaeological and architectural research, along with an attempt at reconstructing the oldest developmental phases of Sandomierz's Dominican Convent, was based on Jerzy Topolski's original methodology of historical knowledge based on building a sequence of models that depict a hypothetical, yet highly probable historical reality. This methodology allowed architects to build a theoretical 3D model of the south wing's reconstruction with dedicated zones for the exhibition of the first layout's relics. This model, as presented in this publication, is aimed at directing the development programme of the building's user and support administrative decision-making by conservation services with academic sources and innovative construction solutions, guaranteeing that the original historical substance will be completely safe, in addition to enabling public access to said substance.

The conceptual proposal of the footing of the monastery's southern wing that is slated for reconstruction, in the form of bored micropiles with a maximum diameter of 100–150 mm and placed at points outside of the outline historical walls will not damage the historical building substance identified by archaeological studies and their interference with archaeological stratigraphy will be minimal. This will allow the construction of the light frame construction of a two-storey (ground floor + first floor and attic, without a cellar) exhibition and museum pavilion, which will initially be used to secure the archaeological site during its exploration. Once this exploration is completed, the building will be arranged in accordance with its intentional form of use. The application of the flexible conservation design method [21] was the result of many years of practical experience of the authors, said experiences being compliant with the fundamentals of contemporary architectural and urban heritage protection theory [22].

References

1. Płuska I (2009) 800 lat cegielnictwa na ziemiach polskich—rozwój historyczny w aspekcie technologicznym i estetycznym (800 years of brickmaking in Poland—historic development in its technological and aesthetic aspect). Wiadomości Konserwatorskie (Conserv News) 26:27
2. Sztuka Polska przedromańska i romańska do schyłku XIII wieku (1969) T. 1, red. M. Walickiego. Wyd. PWN, Warszawa
3. Świechowski Z (1990) Sztuka romańska w Polsce. Arkady, Warszawa
4. Buko A (1998) Początki Sandomierza, IAiEP PAN, IA UW. Zakład Wydawniczy Letter Qality, Warszawa
5. Bugaj U (2016) Wstępne sprawozdanie z badań wykopaliskowych przeprowadzonych w 2016 roku na terenie klasztoru Ojców Dominikanów w Sandomierzu, w otoczeniu tzw. piwnicy o jednym filarze, Warszawa 2016, maszynopis Archiwum ŚWKZ Delegatura w Sandomierzu (in press)
6. Gołembnik A (2019) Badania archeologiczne w rejonie piwnicy o jednym filarze. In: Dominikański klasztor św. Jakuba w Sandomierzu: archeologia i architektura, historia i współczesność. Instytut Historii Architektury i Konserwacji Zabytków na Wydziale Architektury Politechniki Krakowskiej & Instytut Statystyki Kościoła Katolickiego SAC w Warszawie, Kraków–Warszawa, Wyd.
7. Tazbirowa J (1966) Rola polityczna Iwona Odrowąża. Przegląd Historyczny 57(2)
8. Jan Andrzej Spież OP (2016) Święty Jacek Odrowąż. Wydawnictwo W drodze, Poznań
9. Sroczyńska J (2018) Prezentacja interpretacyjna zabytków architektury w ochronie dziedzictwa kulturowego. Wydawnictwo PK, Kraków
10. Lübbe H (1991) Muzealizacja. O powiązaniu naszej teraźniejszości z przeszłością. In: Estetyka w świecie. Kraków

11. Stala K (2019) Współczesne aspekty projektowania osłon w rezerwatach archeologiczno-architektonicznych. Z zagadnień ochrony i ekspozycji dziedzictwa archeologicznego (Contemporary aspects of designing shelters in archaeological-architectonic reserves. Issues of protection and exhibition of archaeological heritage). Wiadomości Konserwatorskie (J Herit Conserv) 60
12. Przygodzki D (2019) Najnowsze tendencje w kształtowaniu ekspozycji archeologicznej (The latest tendencies in archaeological exhibition design). Wiadomości Konserwatorskie (J Herit Conserv) 60
13. https://en.wikipedia.org/wiki/Vesunna_Gallo-Roman_Museum
14. https://www.archaeologisches-museum-frankfurt.de/
15. Die Neue Altstadt (2018) w: Frankfurter Rundschau Geschichte 8(7)
16. Wamers E (2016) 814 Karl der Große 2014. Archäologische und historische Beiträge zu Pfalzen, Herrschaft und Recht um 800. Schnell & Steiner Verlag
17. Vademecum Konserwatora Zabytków (2015) Międzynarodowe Normy Ochrony Dziedzictwa Kultury, Polski Komitet Narodowy Międzynarodowej Rady Ochrony Zabytków ICOMOS Plac Zamkowy 4, Warszawa
18. Pająk M (2006) Wzmacnianie fundamentów zabytkowych budowli na przykładzie stabilizacji kościoła pod wezwaniem św. Piotra i Pawła w Krakowie. Górnictwo i Geoinżynieria 4
19. Gwizdała K, Kościk P (2007) Wykorzystanie iniekcji strumieniowej do wzmacniania posadowień istniejących obiektów budowlanych. Geoinżynieria. Drogi, Mosty, Tunele 2
20. Jurczakiewicz S, Karczmarczyk S (2011) Stabilizacja posadowienia zabytkowych budynków przy pomocy mikropali na tle doświadczeń praktycznych. Czasopismo Techniczne Politechniki Krakowskiej 108(19, 3-B)
21. Rouba B (2008) Projektowanie konserwatorskie. Ochrona Zabytków 56/1(240):57–78
22. Kadłuczka A (2018) Ochrona dziedzictwa architektury i urbanistyki; doktryny, teoria, praktyka. Wydawnictwo Politechniki Krakowskiej, Kraków

Calculation of Thermal Dynamic Characteristics of the Residential Buildings Living Walls

S. M. Kalinović, J. M. Djoković, R. R. Nikolić, and B. Hadzima

Abstract The energy efficiency of green walls, namely, the way of saving the energy consumption by the dynamic characteristics of the façade walls is presented in this paper. Calculation and analysis of the dynamic characteristics of walls, with and without the plant cover, were performed. Comparison of the dynamic behavior of the facade, with and without the vegetation layer, is given and advantages of the former are emphasized, as well. Three different walls with different structures, but of the same individual thicknesses of layers and the same total thickness of the wall construction, were analyzed. The difference is in the thermal mass layer in the construction of a wall made of concrete with stone aggregates, solid brick, or solid blocks of lightweight concrete. Then, the three different walls of the same composition, as in the previous case, were analyzed, with upgraded layers of the soft wood and ivy representing the living layer of the wall.

Keywords Heat transfer coefficient · Thermal absorption · Periodic heat transfer · Decrement factor · Thermal inertia · Green facades · Living walls

1 Introduction

Majority of the world's population lives in cities today and the tendency towards the city life is increasing yearly. Globally, urbanization represents one of the most important environmental issues. Greenhouse gas emissions, reduction of energy sources and urban heat islands are environmental problems associated with urban density. In densely populated areas, the replacement of green zones by artificial surfaces is one of the main causes of emergence of the urban (city) heat islands, the phenomenon of temperature difference between cities and suburban settlements or villages. As

S. M. Kalinović · J. M. Djoković
Technical Faculty of Bor, University of Belgrade, Vojske Jugoslavije 12, 19210 Bor, Serbia

R. R. Nikolić (✉) · B. Hadzima
Research Center, University of Žilina, Univerzitna 8215/1, 010 25 Žilina, Slovakia
e-mail: ruzicarnikolic@yahoo.com

Z. Zembaty et al. (eds.), *Environmental Challenges in Civil Engineering*, Lecture Notes in Civil Engineering 122,
https://doi.org/10.1007/978-3-030-63879-5_8

cities grow, more and more heat is trapped in them, resulting in an increase in air temperature of 6–8 °C compared to the suburbs [1]. Buildings and sidewalks are made mostly of the non-reflective and waterproof building materials, which accumulate solar radiation during the day and then release the heat at night. Heat is also trapped due to reduction of the green spaces in cities, which leads to a decrease in radiation cover, limited air circulation through the city streets between buildings and high production of waste heat from cooling systems, motor traffic, industrial processes, etc. In order to mitigate the effect of urban heat islands and improve the energy efficiency of buildings, use of the green infrastructure on the roofs and facades of buildings has been increased [2].

Since building envelopes generally remain bare, while surrounding areas around buildings are paved or occupied by other structures and parts of roofs are occupied by some form of construction services, the vertical green systems represent the simplest solution.

The green vertical facade systems have several advantages, such as extending the life of the wall, thermal insulation and reducing the sun absorption. Vertical gardens protect the exterior of buildings and homes from ultraviolet radiation, rain, extreme temperature fluctuations and presence of moisture. Their advantages are reduction of heat load and energy consumption, mitigation of the effect of urban heat islands, reduction of air pollution, sound absorption, improvement and preservation of urban biodiversity and aesthetic contribution to the city landscape. The green vertical systems are divided, according to their structural design and characteristics, into green facades and living walls [3]. The green facades are characterized by climbing plants rooted in soil or pots up to a certain height of the wall. Plants climb onto the facade directly against the wall or indirectly, on a support (wire, mesh, or grille) that is located a short distance from the wall. The living walls are made of panels or geotextile felt, which are fixed to a vertical support or wall construction and which provide support to vegetation formed by different plants. The panels can be of different sizes and types, fixed to the wall, with openings containing substrates and plants. Geotextile felt systems use geotextile felt as a substrate for plants or moss and are fixed directly to the wall [4].

The importance of green facades has been recognized by a large number of researchers. In [5] was shown that vegetation on the outer walls stabilized the indoor air temperature in buildings and reduced heat losses and gains through the outer walls as well, resulting in a decrease in the building energy load. In the work [6], a mathematical model of an outer wall, covered with vegetation, was derived to evaluate the thermal effects of plants on the heat transfer through the facades of buildings. This model allows an analysis of how different physiological parameters of plants, such as leaf surface index, average leaf size and leaf absorption, can improve the thermal characteristics of the facade by reducing the external wall surface temperature and heat flux through the facade. Authors of [7] presented results of a series of experiments performed on ivy overgrown walls and on walls without the ivy. They examined four pairs of walls of different orientations at existing university buildings, during the summer in Chicago, Illinois, USA. The results of the experiment showed that the ivy layers reduced the outside surface temperature by an average of

0.7 °C across all the facades depending on orientation and time of day. The results presented in that paper show that ivy walls can lead to low energy savings during the summer sunshine conditions and achieve additional energy savings by reducing the rate of air infiltration due to vegetation. Researchers in [8] studied the feasibility of applying a dual-faceted green facade to residential buildings in Hong Kong in order to reduce the cooling energy consumption during hot and humid summers. They came to the conclusion that significant energy savings were possible. In the work [9], the results of experimental studies, performed on a vertical green wall in a continental Mediterranean climate, are presented. The effect of insulation thickness on the energy performance of the green wall was analyzed using a new methodology called the Green Facade Optimization (GFO). The results showed that a complex wall structure acts as a passive cooling system when the facade is moderately insulated, up to an insulation thickness of 9 cm, above which, more insulation becomes redundant and inefficient. Authors of [10] have performed a two-year experiment at the University of Bari in Italy to determine whether evergreens were suitable for green facades in the Mediterranean climate. The results presented in that paper make it possible to fill the gap in the literature regarding the lack of data for all the seasons in order to be able to see the complete picture of the thermal characteristics of buildings in the Mediterranean climate region. In [11] different technical and economic aspects of creepers on the façade, when climatic restraint differ, were considered through a case study. In [12] authors have focused on influence of the leaf thickness on thermal characteristics of green facades in hot and humid climates. The measurements were made on the south-facing wall that was covered with green ivy facade systems in Suzhou, China. Their results confirmed the importance of the leaf thickness for the thermal characteristics of the green facade systems.

Considering the importance of green walls, in this paper are analyzed the thermal dynamic characteristics of three different structures of walls, with and without the plant cover, in order to find the best solutions for residential buildings in Serbia. The analysis was limited to living wall analysis, while the green facades will be the subject of further research.

2 Problem Formulation

Equation of the heat conduction through the wall in the direction of the wall thickness, when there is no generation of heat in the wall, for the non-stationary conditions, is:

$$a\frac{\partial^2 T}{\partial x^2} = \frac{\partial T}{\partial \tau}, \tag{1}$$

where variable $a = \lambda/\rho c$ (m^2/s) represents the thermal diffusivity of the material and characterizes the rate of heat diffusion through the given material, c is the specific heat capacity, ρ is density and λ is the thermal conductivity.

Relationship between the complex temperature amplitude and flux at surfaces of a homogeneous single-layer wall, in the matrix form, is [13]:

$$\begin{pmatrix} \hat{T}_{we}(j\omega) \\ \hat{q}_{we}(j\omega) \end{pmatrix} = \begin{bmatrix} Z_{11} & Z_{12} \\ Z_{21} & Z_{22} \end{bmatrix} \begin{pmatrix} \hat{T}_{wi}(j\omega) \\ \hat{q}_{wi}(j\omega) \end{pmatrix}, \tag{2}$$

where $\hat{q}_{wi}$ and $\hat{q}_{we}$ are the thermal fluxes on the inner and outer wall surfaces in the complex domain, while $\hat{T}_{wi}$ and $\hat{T}_{we}$ are the amplitudes of the temperature on the inner and outer wall surfaces in the complex domain, respectively. Here Z_{nm} $(n, m = 1, 2)$ elements of the wall transfer matrix with the characteristics λ, ρ, c and d, which relate temperature and the heat flux on one side of the wall with the same variables on the other side of the wall, take the following form:

$$Z_{11} = Z_{22} = \cosh(\xi + j \cdot \xi) = \cosh(\xi)\cosh(j \cdot \xi) + \sinh(\xi)\sinh(j \cdot \xi) \tag{3}$$

$$\begin{aligned} Z_{12} = -\frac{\delta}{2\lambda}[(\sinh(\xi)\cos(\xi) + \cosh(\xi)\sin(\xi)) + j \cdot (\cosh(\xi)\sin(\xi) \\ - \sinh(\xi)\cos(\xi))] \end{aligned} \tag{4}$$

$$\begin{aligned} Z_{21} = -\frac{\lambda}{\delta}[(\sinh(\xi)\cos(\xi) - \cosh(\xi)\sin(\xi)) + j \cdot (\sinh(\xi)\cos(\xi) \\ + \cosh(\xi)\sin(\xi))] \end{aligned} \tag{5}$$

where:

$$\delta = \sqrt{\frac{\lambda T}{\pi \rho c}} \quad \text{and} \quad \xi = \frac{d}{\delta}, \tag{6}$$

and d is the thickness of the wall.

By system of Eq. (2), elements of the matrix defined by Eqs. (3)–(5), relate the temperatures and heat fluxes, i.e. their corresponding complex amplitudes, on one side of the wall surface with the same variables on the other side of the wall surface, at a frequency ω.

Equation (2) applies only to homogeneous single-layered walls. The buildings' walls are, on the other hand, generally multi-layered and heterogeneous, so the given solutions are not valid. However, a solution for the multi-layered heterogeneous wall can be derived from Eq. (2). Writing the matrix Eq. (2) for each wall layer i $(i = \overline{1, n})$ and assuming that the boundary values between layer i and layer $i + 1$ are equal, the matrix Eq. (2) for a heterogeneous wall with n layers, can be written as:

$$\begin{pmatrix} \hat{T}_{we} \\ \hat{q}_{we} \end{pmatrix} = \begin{bmatrix} Z_{N,11} & Z_{N,12} \\ Z_{N,21} & Z_{N,22} \end{bmatrix} \begin{pmatrix} \hat{T}_{wi} \\ \hat{q}_{wi} \end{pmatrix}, \tag{7}$$

where: n is the number of layers of the wall. For each layer, the thermal and physical parameters λ_i, ρ_i, c_i and thickness d_i are defined. Counting starts from the inside. The product of all the Z_i matrices is $Z_N = Z_n Z_{n-1} \ldots Z_3 Z_2 Z_1$.

The $Z_{N,nm}$ matrix elements, define the thermal dynamic parameters of a wall that describe the most important time-dependent processes, which take place in a wall. Those parameters are: $Y_{11/22}$ (W/m^2 K)—internal and external thermal transmittance, Y_{12} (W/m^2 K)—periodical thermal transmittance, f (–)—decrement factor, $\tau_{11/22}$ (h)—phase delay of thermal transmittance (internal/external), τ_{12} (h)—thermal transmittance phase delay and $k_{1/2}$ (J/m^2 K)—surface heat capacities.

The thermal admittance is a complex variable that represents the ratio of the complex amplitude of the heat flux change on one side of the wall and the complex amplitude of the temperature change on the same side of the wall, when the temperature on the other side of the wall is constant and calculated according to the following equations:

$$Y_{11} = -\frac{Z_{11}}{Z_{12}}, \quad Y_{22} = -\frac{Z_{22}}{Z_{12}}. \tag{8}$$

The periodic thermal transmittance is a complex variable that represents the ratio of the complex amplitude of the heat flux change of on one side of the wall and the complex amplitude of the temperature change on the other side of the wall, when the temperature on the first side of the wall is constant and it is calculated as:

$$Y_{12} = -\frac{1}{Z_{12}}. \tag{9}$$

The decrement factor is a ratio of the periodic thermal transmittance modulus and the stationary thermal transmittance (that is, the heat transfer coefficient), namely

$$f = \frac{|Y_{12}|}{U} \tag{10}$$

Value of the decrement factor is a degree of attenuation of the wall temperature amplitude on the outside with respect to the temperature amplitude on the inside. The value of this factor can be between 0 and 1. If it is closer to 0 that means that the wall has a large damping, which is a characteristics of massive structures.

The phase delay of the periodic thermal transmittance and thermal admittance (internal/external) is a period for which the amplitude of the temperature, i.e. the amplitude of the heat flux, passes through the structural element and they are defined as:

$$\tau_{12} = \frac{T}{2\pi} \arg(-1/Z_{12}), \quad \tau_{11/22} = \frac{T}{2\pi} \arg(Y_{11/22}), \tag{11}$$

respectively.

As with the decrement factor, the lighter structures have a shorter phase delay than the massive ones.

The surface heat capacities represent the heat capacity per unit area of a structural element and they are defined as:

$$k_1 = \frac{T}{2\pi}\left|\frac{Z_{11}-1}{Z_{12}}\right|, \quad k_2 = \frac{T}{2\pi}\left|\frac{Z_{22}-1}{Z_{12}}\right|. \tag{12}$$

Calculating all the above parameters, considering that elements of the Z_{nm} transfer matrix appear in all the equations, requires knowing the thermal characteristics of each wall layer λ_i, ρ_i, c_i, as well as its thickness d_i.

3 Results and Discussion

The procedure for calculating the dynamic characteristics of walls includes: identifying the materials that make the layers of a building component, thickness of those layers and the thermal characteristics of the materials; determination of period of temperature change on surfaces; calculation of the heat penetration depth for the material of each layer; determination of the transfer matrix for each layer (component) and multiplication of the transfer matrix of each layer, excluding those peripheral layers, in the correct order to obtain the transfer matrix of the wall.

Below, three different wall structures of the same total thickness with and without plant cover, were considered.

All the data on wall structures are given in Tables 1, 2 and 3 and Fig. 1. Table 1 shows dimensions of individual layers, constituting different wall structures, with the same total thickness. Table 2 presents composition of different walls. In Table 3 are presented characteristics of different materials applied in considered walls. Figure 1 shows thickness, heat conduction coefficient and specific mass of walls 1–6.

Table 1 Wall structure—dimensions of layers

Wall structure	Layer thickness of the wall (m)					
	Wall 1	Wall 2	Wall 3	Wall 4	Wall 5	Wall 6
Extension lime mortar	0.012	0.012	0.012	0.012	0.012	0.012
Concrete with stone aggregates	0.2			0.2		
Full brick		0.2			0.2	
Lightweight concrete blocks			0.2			0.2
Thermal insulation-stone wool	0.08	0.08	0.08	0.08	0.08	0.08
Synthetic board made of multilayer polyester	0.1	0.1	0.1	0.1	0.1	0.1
Softwood panels	–	–	–	0.015	0.015	0.015
Ivy	–	–	–	0.2	0.2	0.2

Table 2 Wall structure—composition of layers

#	Structure	
1		extension lime mortar concrete with stone aggregates thermal insulation - stone wool synthetic board made of multilayer polyester
2		extension lime mortar brick thermal insulation - stone wool synthetic board made of multilayer polyester
3		extension lime mortar lightweight concrete block thermal insulation - stone wool synthetic board made of multilayer polyester
4		extension lime mortar concrete with stone aggregates thermal insulation - stone wool synthetic board made of multilayer polyester softwood panel plants
5		extension lime mortar brick thermal insulation - stone wool synthetic board made of multilayer polyester softwood panel plants
6	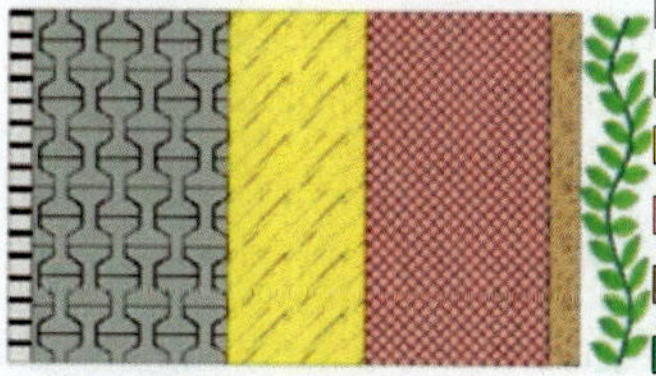	extension lime mortar lightweight concrete block thermal insulation - stone wool synthetic board made of multilayer polyester softwood panel plants

Table 3 Characteristics of the wall materials

Material	ρ (kg/m^3)	c (J/kg K)	λ (W/m K)
Extension lime mortar	1700	1050	0.850
Concrete with stone aggregates	2200	960	1.510
Full brick	1400	920	0.58
Lightweight concrete blocks	1000	840	0.47
Thermal insulation-stone wool	30	840	0.038
Synthetic board made of multilayer polyester	1400	1590	0.190
Softwood panels	110	1880	0.14
Ivy	533	2.8	0.4

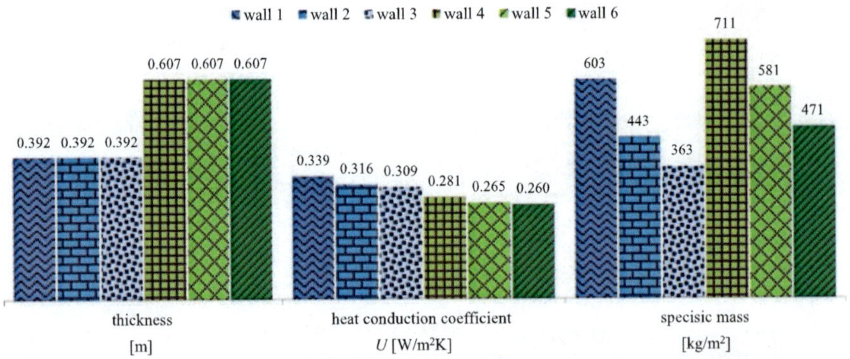

Fig. 1 Characteristics of walls 1–6

Based on results shown in Fig. 1, it can be observed that the difference in the heat transfer coefficients between the walls with and without the plant cover is approximately the same for all the three wall types. The specific mass has a similar trend but in the opposite direction.

The dynamic characteristics of the walls, determined based on Eqs. (3)–(12), are given in Fig. 2. Figure 2 shows internal and external thermal admittance, periodic thermal transmittance and thermal capacity of the inner and outer wall surfaces, decrement factor and the temperature delay factor for walls 1–3 shown in Table 2.

Based on results shown in Fig. 2, it can be seen that the phase delay on the inside of the wall, i.e. internal admittance, is different for different types of walls, while the phase delay on the exterior of the wall, i.e. external admittance, is equal for all the walls. This means that the walls respond to changes in outside temperature at different rates.

The heat capacity of the wall surface on the outside has the same value for all the three types of walls, while the heat capacity on the inside has the highest value for wall 1 and slightly lower values for walls 2 and 3.

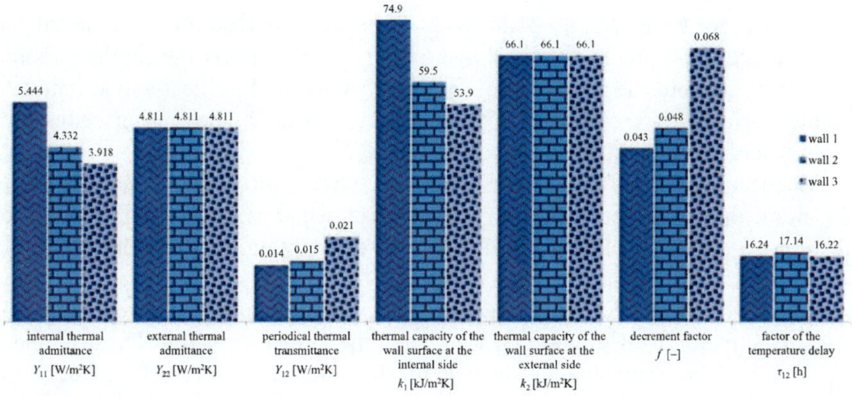

Fig. 2 Dynamic thermal characteristics of walls 1–3

The factor of the amplitude decrease of the external temperature oscillation, that is, the decrement factor, is the highest for wall 3 and lowest for wall 1.

The delay factor for variation of the outside temperature on wall 1 is slightly different from the same value for wall 3, while for wall 2 the delay factor is slightly higher.

The dynamic characteristics for walls 4–6, determined based on Eqs. (3)–(12), are given in Fig. 3. Figure 3 shows internal and external thermal admittance, periodic thermal transmittance and thermal capacity of the inner and outer wall surfaces, the decrement factor and the temperature delay factor for walls 4–6.

Based on results shown in Fig. 3, it can be seen that the phase delay on the inside of the wall, i.e. internal admittance, is different for different types of walls, while the phase delay on the exterior of the wall, i.e. the external admittance, is equal for all the walls. This means that the walls respond to changes on outside temperature at different rates.

When values of the phase shift on the inside of the wall, for walls with and without vegetation cover are compared, it can be seen that those values are identical. On the other hand, values for the phase delay on the outside of the wall are almost four

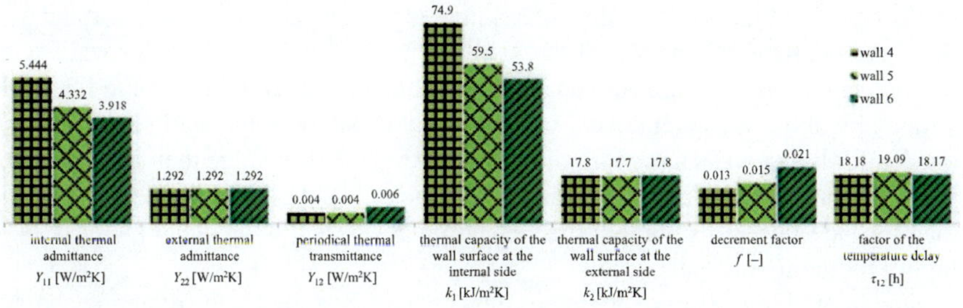

Fig. 3 Dynamic thermal characteristics of walls 4–6

times lower for walls with the plant cover. That means that the ability of the outside wall surface to accumulate thermal energy is four times lower during the periodic variation of the outside temperature, i.e. a small amount of the heat will accumulate on the outside surface of the wall, while a larger amount will be transferred through other layers to the interior.

The heat capacity of the outside wall surface has approximately the same value for all the three types of walls with a plant cover, which is almost four times lower than in the case of walls without it. The heat capacity on the inside has the same value as for walls without the plant cover.

The factor of the external temperature oscillation amplitude decrease, i.e. the decrement factor, is the highest on wall 6 and the lowest on wall 4. The decrement factor is 30% lower for living walls than for the same walls without the plant cover.

The delay factor for changing the outside temperature on wall 4 is slightly different from the same quantity for wall 6, while for wall 5 the delay factor is slightly higher. A comparison of values, shown in Figs. 2 and 3, shows that the delay factor in variation of the outside temperature is 11% higher for the living walls. Based on this, it can be concluded that the walls with the plant cover are walls with greater possibility of absorption of the thermal energy; the thermal mass of the walls is greater and the longer they emit the heat towards the interior, the more stable the change of thermal comfort (the change will be more moderate, with little chance for appearance of peaks and sudden changes of the interior temperatures).

4 Conclusions

The paper analyzes the ability to temporarily delay the external thermal effect, together with the factors of reducing the extreme external temperatures on the inside of the wall in the living walls. It is shown that the achievement of the corresponding values of the wall heat transfer coefficient is related to its structure.

In order to find the optimal solution for the wall structure, with the highest energy saving potential, three types of walls, with different structure and the same total thickness, with and without the plant cover, were analyzed.

Based on results presented in this paper, it can be concluded that the phase delay on the outside of the wall, i.e. the outer admittance, is equal for all the walls and that the phase delay on the outside of the wall is almost four times lower for walls with the vegetation cover. The heat capacity on the inside has the same value for walls with and without the plant cover, while the heat capacity of the wall surface on the outside for living walls has approximately four times lower value than in the case of walls without the plant cover. The decrement factor is 30% lower for the living walls than for the same walls without the vegetation cover, while the delay factor is 11% higher for the living walls.

Due to of all these, more attention must be paid to studying the dynamic characteristics of certain types of walls and to trying to find the best solutions that would meet the prescribed regulations and requirements should be an objective, as well.

Those solutions would, on one hand, be economically viable and on the other, would provide the greatest comfort for housing in buildings.

Acknowledgements This research was partially financially supported by the project "Innovative Solutions for Propulsion, Power and Safety Components of Transport Vehicles" ITMS 313011V334 of the Operational Program Integrated Infrastructure 2014–2020 and co-funded by the European Regional Development Fund and by the Serbian Ministry of Education, Science and Technological Development through grant ON174001 "Dynamics of hybrid systems with complex structures. Mechanics of materials".

References

1. Santamouris M (2012) Sol Energy 103:682–703
2. Alexandria E, Jones P (2008) Build Environ 43:480–493
3. Besir AB, Cuce E (2018) Renew Sustain Energy Rev 82:915–939
4. Riley (2017) Build Environ 114:219–232
5. Yoshimi J, Altan H (2011) In: Proceedings of building simulation 2011, Sydney, 14–16 Nov, pp 1438–1443
6. Susorova M, Angulo P, Bahrami B (2013) Stephens. Build Environ 67:1–13
7. Susorova P, Azimi B (2014) Stephens. Build Environ 76:113–124
8. Wong A (2016) Baldwin. Build Environ 97:34–39
9. Olivieri F, Cocci Grifoni R, Redondas D, Sánchez-Reséndiz JA, Tascini S (2017) Energy Build 150:132–148
10. Vox G, Blanco I, Schettini E (2018) Build Environ 129:154–166
11. Grabowiecki A, Jaworski T, Niewczas A (2019) Belleri. Energy Proc 111:377–386
12. Li C, Wei J, Li C (2019) Energy Build 199:72–87
13. EN ISO 13786/2007

The Durability of Concrete with the Participation of Hydrophilic and Hydrophobic Nanosilica Without and Within the Presence of Silica Fume and New Generation Superplasticizer

E. Janowska-Renkas and D. Matyjaszczyk

Abstract In the paper the research was undertaken in order to determine the impact of hydrophilic nanosilica used in a form of aqueous colloidal solution with admixture of nano-silver, as well as hydrophobic nanosilica dissolved in ethanol and isopropanol on properties of the concrete without and with presence of silica fume and polycarboxylate based superplasticizer. Tests were conducted for concrete mixture (slump test, Vebe test, density, air content by the water column method) and hardened concretes (compressive strength after 1, 7, 14 and 28 days of curing, density, depth of penetration of water under pressure, water absorption, frost resistance). For mixture and hardened concrete without participation of silica fume, Portland cement 42.5 R was used in an amount of 549.7 kg/m^3 for obtaining BWW high-quality concretes. Nano-additives were used in the amount of 0.5 and 1.5% in relation to the mass of cement. The silica fume was additionally added to concretes in the amount of 8% by mass. While as a chemical admixture in the amount of 2% in relation to the cement mass, a 40% solution of a superplasticizer based on a polycarboxylic ether (SP) was used. Pebble aggregate consisting of two fractions with a grain size of 2–8 and 8–16 mm was used and sand with a grain size of 2 mm. Properties of fresh concrete and hardened concrete were tested in the accredited Construction Material Laboratory of the Faculty of Civil Engineering and Architecture at the Opole University of Technology. Whereas the increase of nanoadditive amount to 1.5% by mass contributed to better improvement of concrete physical properties in the presence of hydrophobic nanosilica than in the presence of hydrophilic nanosilica. In that case particularly beneficial properties were demonstrated by "nano-concrete", which apart from nanosilica also contained the silica fume and the superplasticizer, which to the most extent had the impact on the improvement of concrete properties, including its reduced absorptivity, higher compressive strength or higher freeze-thaw resistance compared to other concretes tested.

Keywords Concrete · Durability · Hydrophilic nanosilica · Hydrophobic nanosilica · Silica fume · Chemical admixture

---

E. Janowska-Renkas (✉) · D. Matyjaszczyk
Civil Engineering and Architecture Faculty, Opole University of Technology, Opole, Poland
e-mail: e.janowska_renkas@interia.eu

Z. Zembaty et al. (eds.), *Environmental Challenges in Civil Engineering*, Lecture Notes in Civil Engineering 122,
https://doi.org/10.1007/978-3-030-63879-5_9

1 Introduction

Modification of the concrete started in the 90s of the twentieth century included the following aspects: structural, functional and ecological ones. One of examples for structural modifications are concretes with increased durability. These concretes, unlike the ordinary concrete, have a dense microstructure and a low water-to-cement ratio (w/c ~ 0.2) [1–3]. The most important properties of the concrete currently produced are e.g. very good workability of the fresh concrete, high compressive strength after 28 days—above 60 MPa, water-proof properties, frost resistance, resistance to abrasion and impact of environmental conditions, thermal inertia, sound-proof properties, fire and water protection. Furthermore, it is the ecological material and thus it is not hazardous to human health [1–4].

Achievement of properties mentioned above was possible thanks to modification of the concrete composition, which nowadays is a multi-constituent composite material composed of cement, aggregate, water, chemical admixtures, including effective superplasticizers, as well as the waste substances in a form of mineral additives that show pozzolanic and hydraulic properties, such as e.g.: silica fume, fly ash or blast furnace slag [1–3].

Since the beginning of the twenty-first century, a quick development of nanomaterials application in the concrete technology has been observed. As research [4–7] showed also in case of the concrete, a beneficial way to increase its utility parameters is to introduce nanoparticles to the cement or the concrete. Until now, the impact of such nanoparticles as the oxides of: TiO_2, Fe_2O_3, SiO_2, Al_2O_3 as well as nanoparticles of silver and copper have been best tested. And thus e.g. TiO_2 in the nanoform has the impact on achievement of self-cleaning properties by the concrete, and introduction of Ag and Cu nanoparticles—of bactericidal properties. Whereas introduction of nano—Fe_2O_3 or SiO_2 causes the increase of the flexural and compressive strength of cement materials [4, 8–11].

Studies of many authors demonstrated that introduction of a small amount of ZnO, Cr_2O_3 nanoparticles, from 0.5 to 2% by mass, causes only a slight improvement of mechanical properties; the growth in strength was generally from 5 to 20% [9, 12], a slightly higher in case of nano SiO_2 and Al_2O_3 introduction—from 16 to 45% [9, 13], whereas introduction of nano ZnO and TiO_2 even caused deterioration of the compressive strength by as many about 30% [14, 15], and in the case of nano TiO_2 even the stop of hydratation [15]. Nano-modification in this case is based on introduction of the following nano-oxides to the concrete mixture: iron, aluminium, titanium, silica oxides or even carbon nanotubes or nanosilica with particle size approximately 5 nm [16–21].

Spectacular researches are attributable to Iranian engineers, an example of which is a Persian nanoconcrete proposed by professor Nazari [19]. They demonstrate that inclusion of nanoparticles of iron, aluminium, zirconium, titanium or copper nano-oxides to the ultra-high performance concrete allows to increase its impact strength, which is 4 times higher than the impact strength of Ductal concrete, which makes it more attractive for military applications: to build fortifications and bunkers.

In the paper the research was undertaken in order to determine the impact of hydrophilic nanosilica used in a form of aqueous colloidal solution with admixture of nano-silver, as well as hydrophobic nanosilica dissolved in ethanol and isopropanol on properties of the concrete without and with presence of silica fume and polycarboxylate based superplasticizer.

2 Experimental Part

2.1 *Material*

Composition of designed concretes is given in Table 1.

Tests were conducted for concrete mixture and hardened concretes of Portland cement—CEM I 42, 5R with and withput the water solution of hydrophilic nanosilica colloid with the admixture of nano-silver and hydrophobic nanosilica colloid dissolved in ethanol and isopropanol. For mixture and hardened concrete without participation of silica fume, Portland cement 42.5 R was used in an amount of 549.7 kg/m^3 for obtaining BWW high-quality concretes. Nano-additives were used in the amount of 0.5 and 1.5% in relation to the mass of cement. The silica fume was additionally added to concretes in the amount of 8% by mass. While as a chemical admixture in the amount of 2% in relation to the cement mass, a 40% solution of a superplasticizer based on a polycarboxylic ether (SP) was used. Pebble aggregate consisting of two fractions with a grain size of 2–8 and 8–16 mm was used and sand with a grain size of 2 mm.

2.2 *Test Methods*

Tests included determination of concrete mixtures properties in scope of: consistency: slump test according to PN-EN 12350-2:2011 and Vebe test according to PN-EN 12350-3:2011; density according to PN-EN12350-6:2011; air content by the water column method according to PN-EN 12350-7:2011. Whereas tests of hardened concretes were conducted to determine: compressive strength after 1, 7, 14 and 28 days of curing according to PN-EN 12390-3:200, density according to PN-EN 12390-7:2001; depth of penetration of water under pressure according to PN-EN 12390-8:2001; water absorption according to PN-B-06250:1988 and frost resistance according to PN-B-06250:1988, determined after 28-day curing of concretes.

Properties of fresh concrete and hardened concrete were tested in the accredited Construction Material Laboratory of the Faculty of Civil Engineering and Architecture at the Opole University of Technology.

Table 1 Composition of concretes with and without the addition of hydrophilic and hydrophobic nanosilica, as well as silica fume and polycarboxylate superplasticizer

Identification of concrete	Concrete composition	Cement (kg/m^3)	Aggregate (kg/m^3)	Water (kg/m^3)
M1	100% C	549.7	1619.6	211.4
M2	C+8% D	467.3		
M3	C+0.5% HI	549.7		208.7
M4	C+0.5% HF			
M5	C+0.5% HI+8% D	467.3		209.1
M6	C+0.5% HF+8% D			
M7	C+0.5% HI+8% D+0.4% SP			207.2
M8	C+0.5% HF+8% D+0.4% SP			
M9	C+1.5% HI	549.7		203.2
M10	C+1.5% HF			
M11	C+1.5% HI+8% D	467.3		204.4
M12	C+1.5% HF+8% D			
M13	C+1.5% HI+8% D+0.4% SP			202.5
M14	C+1.5% HF+8% D+0.4% SP			
Where: C-CEM I 42.5R Portland cement	HI-hydrophilic nanosilica HF-hydrophobic nanosilica	D-silica fume SP-polycarboxylate superplasticizer		

Specimens for testing were prepared in line with normative acts (PN-EN standards) taking into account a number of test specimens, as well as their dimensions and size.

Due to the number of test specimens, an error of results obtained was estimated based on the expanded uncertainty of the measurement determined for test methods applied in the study, in line with requirements imposed by the Polish Centre for Accreditation for testing laboratories. Results of tests presented in the article are averaged values. And so conformity of test results was determined for the concrete mixes in scope of:

- testing of consistency: with a slump-test according. to PN-EN 12350-2:2011—taking into account the expanded uncertainty of the measurement equal to ± 1 cm,

whereas for the Vebe test according to PN-EN 12350-3:2011—the expanded uncertainty of the measurement equal to ±1.5 s;
- testing of density according to PN-EN 12350-6:2011—taking into account the expanded uncertainty of the measurement equal to 10 kg/m^3;
- testing of the air content with the water column method according to PN-EN 12350-7:2011—taking into account the expanded uncertainty of the measurement equal to ±0.5%.

Testing of hardened concretes according to PN-EN 12390-3:200 was performed on cubic test specimens with dimensions 100 × 100 × 100 mm. Obtained results of compressive strength tests on test specimens of CEM I 42.5 R cement without additives, as well as modified concretes with nanoadditives with and without silica fume and plasticizer after 1, 7, 14 and 28 days of curing, taking into account the expanded uncertainty of measurement equal to ±2.9 MPa. Then subsequently in tests of density according. to PN-EN 12390-7:2001, the expanded uncertainty of measurement taken into account was ±8.7 kg/m^3, in tests of depth of penetration of water under pressure according to PN-EN 12390-8:2001—the expanded uncertainty was ±4 mm, and in tests of water absorptivity according to PN-B-06250:1988—the expanded uncertainty was ±0.36%. Whereas in freeze-thaw resistance tests according to PN-B-06250:1988—the expanded uncertainty equal to ±0.46%.

Pursuant to requirements of relevant standards, a number of test specimens in individual tests depended on the testing method applied and it was respectively:

- for testing of consistence with the Vebe method according to PN-EN 12350-3:2011 and the slump test. PN-EN 12350-2:2011, according to PN-EN 12350-6:2011 and the air content in concrete with the water column method according. to PN-EN 12350-7:2011 measurements were conducted for each concrete batch mixed (MB) without and with modification of the composition with the content of nanoadditives, silica fume and the superplasticizer introduced according to the composition given in Table 1
- for compressive strength tests according to PN-EN 12390-3:2001 measurements were performed on three test specimens of the same type of concrete, and the result was averaged in line with guidelines set forth in the standard. Tests were repeated twice.
- tests of concrete freeze-thaw resistance according. to PN-88/B-06250 were performed for 12 test specimens of the same type of concrete, whereas 6 specimens were subject to freeze-thaw tests, and remaining 6 to curing in standard conditions in temperature 20 ± 2 °C), while treating them as witness specimens in line with requirements of the standard applied. After a 28-day period, test specimens were subject to the compressive strength testing.

3 Results of Testing and Their Analysis

Analysis of test results obtained showed that application of hydrophilic and hydrophobic nanosilica without and with participation of silica fume and polycarboxylate superplasticizer had the impact on properties of both fresh concrete and hardened concrete. Therefore amodification of the test (reference) concrete mixture (M1) with hydrophilic and hydrophobic nanosilica content used in amounts of: 0.5% and 1.5% by mass (in mixtures: M3, M4, M9, M10), affected the cone slump increase about from 3 to 6 cm, which caused a change of consistency class from S3 (determined for the M1 test mix) to S4. Also the concrete mixtures containing 8% by mass of silica fume and 0.4% by mass of superplasticizer SP (M7, M8, M13 and M14) had the S4 consistency class. Whereas concrete mixtures that apart from the silica fume contained the HI hydrophilic nanosilica (M5 and M11) or HF hydrophobic nanosilica (M6 and M12), used in the amount of 0.5 and 1.5% by mass, showed the reverse effect, i.e. the slump reduction by ca. 4–7 cm compared to the test mixture, and thus the change of the consistency class from S3 class to S2 class.

For concrete mixes containing nanosilica, the vibration time determined by means of the Vebe method, was shorter than 5 s, which proves the lack of reliable results and excludes application of this method to determine consistency of the above concrete mixes (Table 1).

The admixture of both hydrophilic and hydrophobic silica to the concrete mix without and with the silica fume and superplasticizer (M3-M14) increased in general the density of mixtures tested compared to the density of the test specimen of pure CEM I 42.5R Portland cement (M1). Only the modified mix with 8% by mass of silica fume—M2—had lower density (Table 1).

It was found that with lower content of nanosilica (0.5% by mass) (Table 1), the air content in the concrete mix is lower in the presence of M3 hydrophilic nanosilica (1.5%) than with the M4 hydrophobic nanosilica (1.9%). On the other hand, increase of the nanosilica amount from 0.5 to 1.5% by mass causes a reverse effect demonstrated by higher air content in the mix containing M10 hydrophilic nanosilica (1.9%) than in case of the mixture with M9 hydrophobic nanosilica (1.6%). Increase of air content along with the increase of nanosilica amount was also observed in the paper [15]. It was demonstrated that introduction of silica fume to the concrete mix containing the hydrophobic nanosilica, used in 0.5% by mass, caused reduction air content by over twice in M6 mix (0.8%). Whereas in the presence of polycarboxylate superplasticizer, concrete mixes containing the nanoadmixture and silica fume (M7-M13 and M8-M14) showed increased and comparable air content (ca. 1.7–1.9%) regardless of the nanosilica type (Table 2).

The impact of hydrophilic and hydrophobic nanosilica, used both in the amount of 0.5 and 1.5% by mass, on the density increase of hardened concretes (B3, B9, B4 and B10) was found. Density of those concretes was identical and amounted to 2370 kg/m^3, which compared to density of B1 test concrete (2210 kg/m^3) gave the increase by 160 kg/m^3.

Table 2 Test results of consistency, air content (with Vebe and water column methods) and density of the concrete mix of CEM 42.5R cement without and with the content of hydrophilic and hydrophobic nanosilica, as well as silica fume and superplasticizer

Identification of concrete mix	Composition of concrete mix	Qty of n SiO_2	Consistency				Air content	Density of mix
		(%)	Vebe (s)	Class	Cone slump (cm)	Class	(%)	(kg/m^3)
M1	100% C	–	6	V3	14	S3	1.8	2310
M2	C+8% D	–	6	V3	10	S3	1.6	2290
M3	C+0.5% HI	0.5	2	V4	17.5	S4	1.5	2320
M4	C+0.5% HF	0.5	3	V4	18	S4	1.9	2330
M5	C+0.5% HI+8% D	0.5	4	V4	9.4	S2	1.4	2310
M6	C+0.5% HF+8% D	0.5	7	V3	7.2	S2	0.8	2310
M7	C+0.5% HI+8% D+0.4% SP	0.5	2	V4	17.5	S4	1.9	2330
M8	C+0.5% HF+8% D+0.4% SP	0.5	2	V4	17	S4	1.8	2330
M9	C+1.5% HI	1.5	4	V4	17.5	S4	1.9	2320
M10	C+1.5% HF	1.5	3	V4	17.5	S4	1.6	2320
M11	C+1.5% HI+8% D	1.5	3	V4	7	S2	1.6	2310
M12	C+1.5% HF+8% D	1.5	6	V3	8.5	S2	1.4	2320
M13	C+1.5% HI+8% D+0.4% SP	1.5	4	V4	20	S4	1.7	2330
M14	C+1.5% HF+8% D+0.4% SP	1.5	3	V4	18.5	S4	1.9	2330
Expanded uncertainty of the measurement								
	Consistency tested by the slump-test ±1 (cm)				Air content ±0.5 (%) Density of mix ±10 (kg/m^3)			

While introduction of silica fume and superplasticizer, apart from nanosilica, to the concrete composition caused density reduction of the concrete (B5-B8, B11-B14), by ca. 10–30 kg/m^3, compared to density of concretes with nanosilica content (B3, B9, B4 and B10). Only in case of the concrete containing the silica fume (B2) reduction of its density by ca. 30 kg/m^3 was observed compared to density of the test concrete (B1).

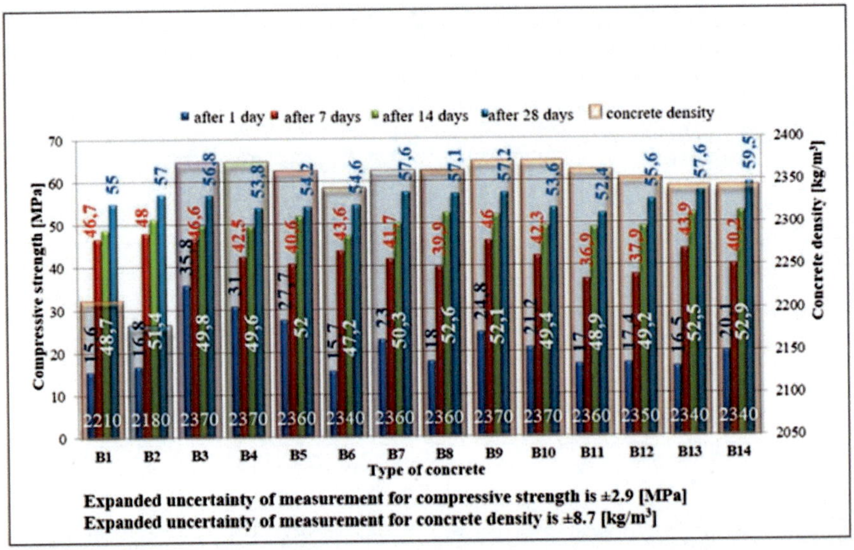

Fig. 1 Compressive strength after 1, 7, 14 and 28 days of concrete curing and density of concrete of CEM I 42.5R without and with addition of hydrophilic and hydrophobic silica and superplasticizer

Modification of concrete composition with the presence of both hydrophilic and hydrophobic silica, as well as silica fume and superplasticizer (B2-B14, Fig. 1) had a beneficial effect on increase of the compressive strength compared to the strength of the test concrete of CEM I 42.5 R Portland cement (B1).

It is interesting that concretes containing a lower (0.5% by mass) amount of hydrophilic silica (B3) showed the highest strength gain observed already after 1 day of curing (by ca. 222%), a slightly lower in the presence (0.5% by mass) of the hydrophobic nanosilica in B4 concrete (by 194%), compared to the test concrete (B1). Increase of the nanosilica content to 1.5% by mass leads to lower gain of early strength, i.e. by 126% for B9 concrete in presence of hydrophilic nanosilica and by 131% for B10 concrete with hydrophobic nanosilica respectively.

From 14 to 28 days, the compressive strength increases for concretes containing the increased (up to 1.5% by mass) amount of nanosilica.

Analyzing the test results it may be concluded that in the initial curing time (after 1 day) dosing of smaller amounts of nanosilica as the concrete constituent has more beneficial impact on early strength gain, other than after a longer, 28-day curing of concretes, for which the strength is higher with the amount of nanoadditive is increased to 1.5% by mass (Fig. 1).

Presence of silica fume and superplasticizer in concretes containing hydrophilic silica (B5, B7, B11, B13) and hydrophobic silica (B6, B8, B12, B14) it had a beneficial effect on strength growth only after 14 days of concrete curing reaching the strength on the level comparable to the concrete with hydrophilic nanosilica (B3, B9) and hydrophobic silica (B4, B10) or even exceeding that value (Fig. 1). Whereas

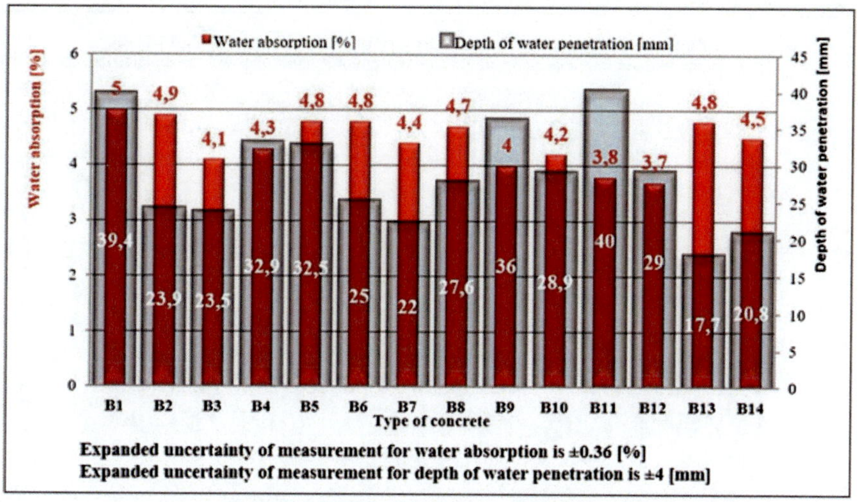

Fig. 2 Water absorption and depth of water penetration into the concrete under pressure of CEM I 42.5R without and with the content of hydrophilic and hydrophobic nanosilica and superplasticizer

after 28 days of curing, the highest compressive strength among the concrete types tested, was demonstrated by the concrete with the following composition: 1.5% by mass of hydrophobic nanosilica, 8% by mass of silica fume and 0.4% by mass of superplasticizer (B14), which indicates beneficial cooperation of selected quantities and types of constituents.

Concretes with addition of both hydrophilic and hydrophobic nanosilica demonstrated similar water absorption values (Fig. 2), which were much lower than water absorption of B1 test concrete (5%).

For other concretes (B2–B14) the additive of hydrophilic and hydrophobic nanosilica, silica fume and superplasticizer had a beneficial impact on reduction of water absorption, below 5%, which allows for their application in elements exposed to direct impact of weather conditions (according to PN-88/B-06250). The expanded uncertainty for the water absorptivity test is ±0.36%.

Addition of silica fume and superplasticizer to the concrete containing hydrophilic and hydrophobic nanosilica had the impact on increase of concrete water-proof (Fig. 3), which is especially visible with the increase dose of nanoadditive. The above is confirmed by the lowest depth of water penetration under pressure to these concretes B7, B8 and B13, B14. According to standard PN-EN 12390-8: 2001, out of 6 examined concrete cubes with dimensions 150 × 150 mm, the result of the test is the largest depth of water penetration into one of the tested concrete samples.

Concretes containing hydrophilic and hydrophobic nanosilica without and with the silica fume and superplasticizer showed higher resistance to frost determined both by the mass loss and strength reduction compared to the test concrete. Samples of these concrete subject to freezing did not show any cracks or visible wastage, and the average mass loss did not exceed 5% compared to test specimen. Those concrete

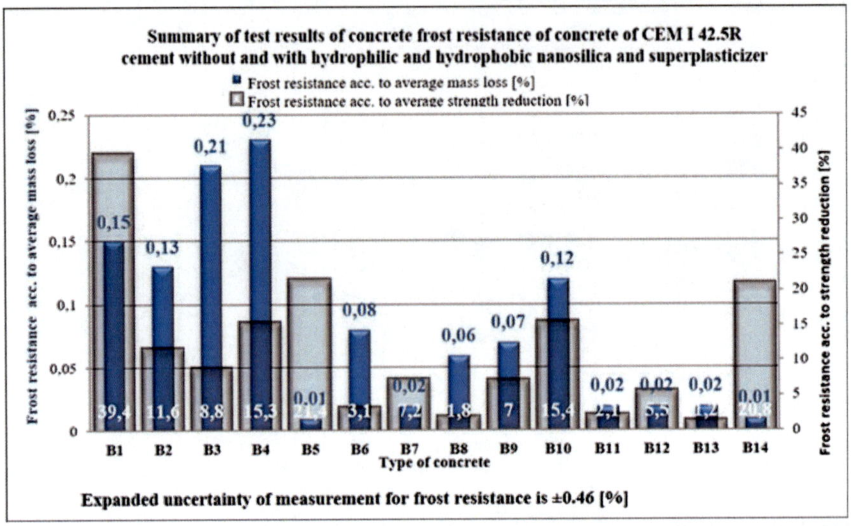

Fig. 3 Frost resistance of concrete of CEM I 42.5R cement without and with the content of hydrophilic and hydrophobic nanosilica and superplasticizer

types reached F50 frost resistance level. Only in case of the concrete containing 0.5 wt% of hydrophobic nanosilica (B4) the required level of frost resistance (F50) could not be reached.

4 Summary and Conclusions

Based on test results obtained it was demonstrated that:

Introduction of nanoadditive to concrete had a beneficial effect on improvement of their workability and reduction of the air content, depending on the type and quantity of the nanoadditive used. And thus the air content in the concrete mix was lower in the presence of M3 hydrophilic nanosilica (0.5%) than with the M4 hydrophobic nanosilica (1.9%). On the other hand, increase of the nanosilica amount from 0.5 to 1.5% by mass caused a reverse effect demonstrated by higher air content in the mix containing M10 hydrophilic nanosilica (1.9%) than in case of the mixture with M9 hydrophobic nanosilica (1.6%).

It was demonstrated that introduction of additives to concretes (B2-B14), regardless of their type and quantity, caused the increase of the density in concretes tested compared to the density of B1 test concrete.

Results of compressive strength tests obtained in the initial period of curing (after 1 day of curing) showed that application of hydrophilic and hydrophobic nanosilica in the amount of 0.5% by mass caused over twice growth of strength compared to the test concrete. With the smaller amount of the nanoadditive (0.5% by mass), the concrete containing hydrophilic nanosilica showed the better strength. Whereas after a longer time (28 days of testing) with the smaller amount of nanoadditive (0.5% by mass), the improvement of concrete application parameters was observed in the presence of hydrophilic nanosilica.

Whereas the increase of nanoadditive amount to 1.5% by mass contributed to better improvement of concrete physical properties in the presence of hydrophobic nanosilica than in the presence of hydrophilic nanosilica. In that case particularly beneficial properties were demonstrated by"nano-concrete", which apart from nanosilica also contained the silica fume and the superplasticizer, which to the most extent had the impact on the improvement of concrete properties, including its reduced absorptivity, higher compressive strength or higher freeze-thaw resistance compared to other concretes tested.

It was found that the presence of hydrophilic and hydrophobic nanosilica could have a significant impact on a change of concrete mixture and hardened concrete properties leading to increasing of its durability depending on the type and quantity of nanoadditive used.

Test results obtained indicate a need of thorough testing of hydrophilic and hydrophobic nanosilica impact on properties of the concrete mix and hardened concrete aiming at reduction of w/c ratio and to investigate the impact of higher content of nanosilica on the structure and tightness of concrete.

References

1. Czarnecki L, Kurdowski W (2006) Tendencies shaping the future of concrete. In: Conference materials „Dni Betonu", Wisła. pp 47–64
2. Raczkiewicz W (2012) Concrete—a construction material known for centuries. Przegląd Budowlany 83:13–18
3. Aïtcin P-C (2000) Cements of yesterday and today: concrete of tomorrow. Cem Concr Res 30:1349–1359
4. Czarnecki L (2011) Nanotechnology in construction. Przegląd Budowlany 82:40–53
5. Radomski W (1995) Global development trends of concrete technology. Przegląd Budowlany 8–9:23–29
6. Kurdowski W (2010) Cement and concrete chemistry. Wydawnictwo Polski Cement, Wydawnictwo Naukowe PWN, Warszawa, p 2010
7. Czarnecki L, Justnes H (2012) Sutainable and durable concrete. Cement Wapno Beton 17:341–362
8. Sobolev K, Gutiérrez MF (2005) How nanotechnology can change the concrete world. Am Ceram Soc Bull 84:14
9. Flores-Velez LM, Dominguez O (2002) Characterization and properties of Portland cement composites incorporating zinc-iron oxide nanoparticles. J Mater Sci 37:983–988
10. Błaszczyński TZ (2012) Nanotechnologies in construction. In: Seminar "Nanotechnologies in construction", international fair Budma, Poznań, pp 1–41
11. Nazari A, Riahi S, Shamekhi SF, Khademno A (2010) The effects of incorporation Fe_2O_3 nanoparticles on tensile and flexural strength of concrete. J Am Sci 6(4):90–93
12. Yang J, Mohseni E, Behforouz B, Khotbehsara MM (2015) An experimental investigation into the effects of Cr_2O_3 and ZnO_2 nanoparticles on the mechanical properties and durability of self-compacting mortar. Int J Mater Res 106:886–892

13. Nazari A, Riahi S, Riahi S, Shamekhi SF, Khademno A (2010) Influence of Al_2O_3 nanoparticles on the compressive strength and workability of blended concrete. J Am Sci 6:6–9
14. Xiong G, XuL Deng M, Tang M (2006) Properties of cement-based composites by doping nano-TiO_2. J Chin Ceram Soc 34:1158
15. Nazari A, Sh R, Sh R, Shamekhi SF, Khademno A (2010) Assessment of the effects of the cement paste composite in presence TiO_2 nanoparticles. J Am Sci 6:43–46
16. Cwirzen A, Habermehl-Cwirzen K, Penttala V (2008) Surface decoration of carbon nanotubes and mechanical properties of cement/carbon nanotube composites. Adv Cement Res 20:65–73
17. Raki L, Beaudoin J, Alizadeh R, Makar J, Sato T (2010) Cement and concrete nanoscience and nanotechnology. Materials 3:918–942
18. Horszczaruk E, Mijowska E, Cendrowski K, Sikora P (2014) Influence of the new method of nanosilica addition on the mechanical properties of cement mortars. Cement Wapno Beton 18:308–316
19. Nazari A (2011) Nanoparticles in self compacting concrete. VDM
20. Ji T (2005) Preliminary study on the water permeability and microstructure of concrete incorporating nano-SiO_2. Cement Concr Res 35(10):1943–1947
21. Li Z, Wang H, Lu Y, Wang M (2006) Investigations on the preparation and mechanical properties of the nano-alumina reinforced cement composite. Mater Lett 60:356–359

Underwater Concrete—Impact of Fine Particles in Cement on Washout Resistance of the Concrete Mix

S. Grzeszczyk and K. Jurowski

Abstract The study discusses a role of chemical admixtures in creation of underwater concrete properties. Particular attention was paid to anti-washout (viscosity enhancing) admixtures (AWA). Their classification and mechanism of action were presented. Results of own testing were analysed with respect to the impact of mineral additives of different fineness (ground-granulated blast-furnace slag, SiO_2 nanoparticles) and admixtures to the cement on the washout resistance of underwater concrete. A favourable impact of fine particles in the cement on increasing the concrete washout resistance was demonstrated. Taking into account a specific chemical structure of the AWA applied, as well as its mechanism of action, a reason for increasing the underwater concrete washout resistance along with increasing quantity of fine particles in the cement, including nanoparticles, was explained.

Keywords Underwater concrete · Chemical admixtures · Anti-washout admixture · Nanoparticles

1 Introduction

Concrete work technology with application of underwater concrete types generally consists in placing the concrete mix directly through a layer of still or flowing water. It is used in places, where the concrete mix being placed is exposed to the water impact, i.e. during construction and repairs of hydrotechnical structures, bridge abutments, foundations on wetlands, oil platforms, strengthening of wharfs, etc. [1–3].

Underwater concrete technologies have been known and used for years, however, they are continuously developed by using more and more interesting solutions for placing the mixes and by modifying their composition with respect to properties. Today's underwater concrete mix is much different than that used over 50 years ago.

A particular importance for development of underwater concrete technologies had the concrete technology developed and implemented by the Japanese in the 1990s, i.e.

S. Grzeszczyk · K. Jurowski (✉)
Faculty of Civil Engineering and Architecture, Opole University of Technology, Opole, Poland
e-mail: k.jurowski@po.edu.pl

Z. Zembaty et al. (eds.), *Environmental Challenges in Civil Engineering*, Lecture Notes in Civil Engineering 122,
https://doi.org/10.1007/978-3-030-63879-5_10

a new generation of polycarboxylate-based superplasticizers, much more effective, and in result the development of self-compacting concrete technology [4, 5].

A progress in development of underwater concrete technologies is associated with the use of achievements from the self-compacting concrete technology. Modern technologies of underwater concrete work involve meeting specific requirements by the concrete mix. The mix should have proper rheological properties that guarantee flowability of the mixture over time and demonstrate the least washout [6–8]. Therefore, application of the self-compacting concrete technology to obtain underwater concrete, apart from addition of last-generation superplasticizers, requires application of the anti-washout admixture (AWA) preventing the washout of mix constituents under water [9].

The paper presents test results from studies of own authors, aiming at increasing of washout resistance of underwater concrete mixes by increasing amount of fine particles in cement.

2 Role of Chemical Admixtures and Mineral Additives in Shaping Underwater Concrete Properties

During underwater concrete placement, properties of the concrete mix are very important. Its composition has to provide suitable rheological properties, as well as the lowest washout as possible [10]. Application of self-compacting concrete for underwater work became possible after development of not only the new-generation polycarboxylate-based superplasticizers, but also introduction of a new type of AWA.

Viscosity enhancing chemical admixtures form a wide group of polymers. They were divided into five groups according to Ramachadran's classification [11]. Firstly, there are synthetic and natural organic polymers soluble in water, then flocculants soluble in water and organic material emulsions providing ultrafine particles. The next group are large-surface non-organic materials swelling in water. The last group are also large-surface materials increasing the share of fine particles, e.g. fly ash, hydrated lime, diatomaceous earth. On the other hand, Kawai [12] presented AWAs classification according to the type of polymer, dividing them into three groups. These are natural polymers including starch, natural rubbers and plant protein, the next group covers semi-synthetic polymers including starch decomposites and derivatives, while the remaining group are synthetic polymers based on ethylene [13].

It is generally assumed that action of these admixtures consists in "cross-linking" of binder particles by long chains of polymers. Chains of polymers interlace fine particles of the cement, which in consequence increases viscosity of the paste (Fig. 1). Underwater concrete mixes, with enhanced viscosity, have better stability, are capable to fill out elements with dense reinforcement, without symptoms of segregation and

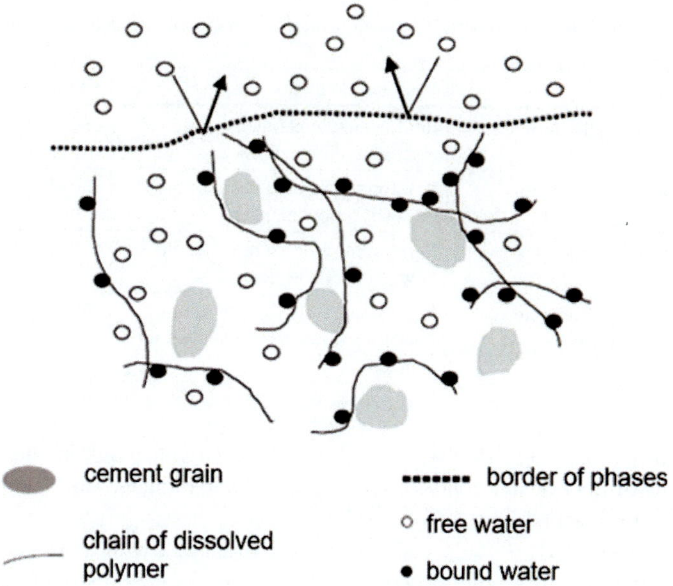

Fig. 1 Mechanism of viscosity enhancing admixture [14]

with minimum blinding, while maintaining suitable flowability provided by addition of the superplasticizer.

A combined application of the superplasticizer with the anti-washout admixture eliminates problems associated with segregation of liquid constituents of concrete mixes, especially during its transport and placement. High cohesion of underwater concrete is provided by its limited contact with water during placement and a small mass loss due to washout. Proper selection of admixtures allow to obtain high flowability of the mixture by application of the superplasticizer that controls the flow distribution (yield stress) and achieve an adequate viscosity that limits the paste washout, which is controlled by the anti-washout admixture.

It is necessary to note that the washout of a underwater concrete mixture depends on many factors. In general, the first to be mentioned is the AWA, its quantity and type. Then, the cement's quantity and composition, water-cement ratio and superplasticiser quantity [9, 15]. The role of fine particles introduced into cement with mineral additives, such as silica fume, finely ground granulated blast-furnace slag and fly ash, in improving the washout resistance of underwater concrete mixes is emphasized [16–19].

Properties of the modern underwater concrete mix are different than those of the traditional mix. It should be self-flowing compacting under its own weight, without vibrating. As its compacting under water is impossible. It should be capable to fill in any shape of formwork, without symptoms of segregation of mix ingredients, but

Table 1 Flow distribution of underwater concrete mixtures, requirements depending on application [20]

Application	Main requirements	Slump-flow d (mm)
Tremie seal for cofferdam	Infill in simple geometry, little obstruction to the flow	280–400
Drilled shafts	Reinforcing steel to obstruct the flow, short flow path	350–450
Reinforced structural walls	Reinforcing steel to obstruct the flow, moderate to flow path	400–580
Reinforced tremie concrete slab with flat top surface	Reinforcing steel to obstruct the flow, long flow path, self-levelling concrete	580–680

first of all demonstrate the washout resistance. American requirements regarding slamp-flow of underwater concrete mixes make this parameter value dependant on application (Table 1).

As data presented in Table 1 show, erected structures with a simple geometry require the mix of lower slump-flow, whereas while placement structural elements that contain dense reinforcing steel, the mix slump-flow should be relatively higher.

3 Material for Testing

To prepare self-flowing underwater concrete mixes, the Portland cement was used CEM I 42.5 R (CEM I), as well as the blast furnace cement CEM III/A 42.5 N-LH/HSR/NA (CEM III). As the mineral additive to the cement CEM I, the ground granulated blast furnace slag was used in various quantity (20%, 50% and 60%) and of different fineness degree marked as C (coarse) and F (fine), with the specific surface area equal to 440 m^2/kg and 540 m^2/kg, respectively.

The particle size analysis of cements (CEM I and CEM III) and slag is presented in Table 2. As tests show, the Portland cement (CEM I) contains much less fine

Table 2 Parameters characterizing fineness of cements (CEM I and CEM III) and slag [21, 22]

Material	Dv (10)	Dv (50)	Dv (90)	Content of particles		
				<5 μm	<10 μm	<20 μm
	μm			(%)		
Cement (CEM I)	6.8	38	68	7	16	32
Cement (CEM III)	2.7	16	49	20	36	59
Coarse slag (C)	4.9	42	59	10	18	36
Fine slag (F)	3.9	17	48	15	31	59

Table 3 Composition of concrete mixes (kg/m^3) [21, 22]

Identification of mix	M1	M2	M	MC20	MC50	MF50	MF60
CEM I			500	400	250	250	200
CEM III	500	500					
GGBFS (C)				100	250		
GGBFS (F)						250	300
Water	170	170	170	170	170	170	170
Aggregate	1680	1680	1680	1680	1680	1680	1680
SP (% mass.)	2	2	2	2	2	2	2
AWA (% mass.)	0.75	0.75	0.75	0.75	0.75	0.75	0.75
Nano SiO_2 (% mass.)		0.5					

particles, below 20 μm, than the slag (C), even though both the cement and the slag have the same specific surface area acc. to Blaine, equal to 440 m^2/kg.

Table 2 presents the parameters characterising the particle size distribution of cements CEM I and CEM III as well as mineral additives used in research. It includes the percentage share of cement particles with diameters smaller than 5 μm, 10 μm and 20 μm, as well as diameters below which there is 10%, 50% and 90% of cement particles (designated accordingly as Dv(10), Dv(50) and Dv(90)). The data presented in Table 2 demonstrates that the greater fineness concerned cement CEM III and fine ground granulated blast-furnace slag. In the case of cement CEM III, 50% of particles has a diameter lower than 16 μm, whereas in additionally ground slag—17 μm. As for cement CEM I it is 38 μm.

To prepare concrete mixes, the natural aggregate was used obtained from mixing of fractions: 0/2, 2/8 and 8/16 at ratios 2:1:1 respectively. For testing of washout, concrete mixes made of CEM III cement were used, without and with addition of SiO_2 nanoparticles, marked as M1 and M2 respectively (Table 3). Concrete mixes were also made, where the following were used as the binder: CEM I 42.5 R (M), mixes where cement (CEM I) with addition of the slag was used in quantity 20% by mass (MC20) and 50% by mass (MC50), as well as cement (CEM I) with addition of blast furnace slag additionally ground in 50% by mass (MSF50) and 60% by mass (MSF60).

SiO_2 nanoparticles were introduced into the M2 concrete mix in amount of 0.5% and 1.0%. Due to a relatively little impact of a higher number of nanoparticles on reduction of mix washout, the additive of 0.5% nanoparticles by mass was used for concrete mix washout testing.

To prepare underwater concrete mixes, the following chemical admixtures were used: polycarboxylate based superplasticizer (SP) in amount of 2% of the cement mass and the cellulose derivatives based AWA in amount of 0.75% of the cement mass.

4 Test Methods

Cement particle size distribution was tested by means of a particle size laser analyzer—Mastersizer 3000 with a wet dispersion method. Isopropyl alcohol was used as a dispersant.

Consistency of the concrete mixture was tested with a slump-flow test method, in accordance with PN-EN 12350-8:2012. The flow diameter (d) and flow time to 500 mm (t_{500}) were determined.

Washout of concrete mixtures was tested in line with CRD-C 61-89A standard.

5 Results and Discussion

5.1 Chemical Structure of Anti-Washout Admixture (AWA) and Its Action

The chemical structure of AWA, defined in the paper [21], based on a Fourier-transform infrared spectroscopy (FTIR) and thermogravimetric analysis, derivative thermogravimetric analysis (TGA-DTG), was used to explain the impact of the blast furnace slag of various fineness and SiO_2 particles added to the cement on the washout of the concrete mix. Tests showed that AWA contains 42.2% of a polymer organic phase and 53.6% of calcium carbonate (chalk) with a small admixture of silicates and/or aluminosilicates.

Based on FTIR and TGA/DTG analyses, it was confirmed that the AWA polymer includes three types of cellulose derivative chains performing various functions in the concrete mixture. These are: methyl cellulose ethers, polyoxyethylene cellulose ethers and carboxy cellulose. The methyl cellulose ethers' chains are responsible for AWAs dissolution in water and formation of colloid which radically increases the medium's viscosity and contributes to water retention in the composition. This takes place thanks to the development of hydrogen bonds between water particles and the oxygen atoms of methoxyl groups. On the other hand, polyoxyethylene cellulose ethers' chains undergo hydration in a water environment and each oxygen atom of the C-O-C ether group can bind two water particles using a hydrogen bond, thereby straightening twisted polyoxyethylene chains in a water environment. Furthermore, polyoxyethylene chains take part in the development of transverse bonds (bridges) between adjacent methyl cellulose chains, thereby creating a dense spatial network constituting a trap for cement paste's fine fractions. This leads to water retention in the composition and increase in environmental viscosity. On the other hand, the COO^- carboxylate groups of sodium salt in a carboxyl cellulose chain are responsible for the electrostatic anchoring of the AWA polymer on particles of fine fractions of the cement paste.

Table 4 Test results for concrete mixes [21]

Mix	Flow diameter d (cm)	Flow time t_{500} (s)	Washout (%)
M1	56	10	2.8
M2	48	–	1.6

A feature common for both chemical additives used, SP and AWA, is a presence of long polyoxyethylene chains in their structure, which decide of their mutual compatibility.

5.2 *Washout of Concrete Mixes*

Results of testing for M1 and M2 concrete mixes are shown in Table 4. Concrete mix M2 containing the addition of nanoparticles shows the lowest slump-flow and clearly higher resistance to washout. Introduction of 0.5% of SiO_2 nanoparticles causes reduction of washout by over 40%. Test results confirmed the relation found in papers [9, 15], i.e. increased resistance to washout of the mix along with reduction of its flowability.

Results of concrete mixes washout tests (M, MC20, MC50, MF50 and MF60) are presented in Fig. 2. The highest washout is demonstrated by the mix (M) made of CEM I cement equal to 2.7%. Addition of the slag reduces the washout significantly, and the more fineness of the added slag and the higher it's content, the higher washout reduction is (Fig. 2).

Tests of particle distribution of blast furnace slag with various specific surface areas demonstrated that the slag with larger surface area contains much more fine particles than the cement. Analysis of fine particles content up to 20 μm is presented in Table 2.

Addition of the slag (C) with a lower content of fine particles to the cement, in the amount of 20% and 50% by mass slightly reduces the concrete mix washout compared to the reference mix, by 0.5% and 0.7% percentage point respectively. A

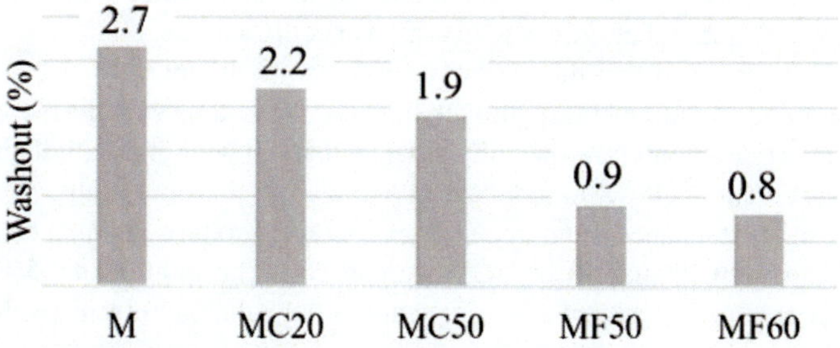

Fig. 2 Washout of concrete mixes [22]

clear improvement of the resistance to washout is observed in concrete mixes, where the cement with addition of the slag (F) with higher specific surface area was applied. This slag contains almost twice as many particles with diameter below 5 μm and almost three times more particles below 10 μm than the slag (C). In this case the washout of the concrete mix with the slag content of 50% by mass is only 0.9%.

Therefore, it may be concluded that the increase of fine particles in the concrete mix, due to introduction of the slag to cement, causes higher resistance to washout, the larger quantity of fine particles is (below 20 μm)—Fig. 2. Comparison of washout of concrete mixes MF50 and MC50, that contain the same amount of the slag (50% by mass), shows that the washout of the mix containing finer slag particles, and thus a higher content of fine particles, is lower by 1.0% point.

Reduction of underwater concrete washout together with the increase of fine particle content in CEM I, caused by addition of the slag and with introduction of nanoparticles, can be explained if we take into account the chemical structure of AWA used, determined in the previous study [21]. This phenomenon is closely related to the chemical structure and mechanism of action of the AWA applied, in the liquid phase environment of the concrete mix. Nanoparticles and fine particles of the blast furnace slag, due to their size, have possibility to get located in the formed dense lattice of developing ether chains of ethylene polyoxide and methyl cellulose chains, due to hydration of ether groups occurring in both chains. It hinders migration of water, cement grains and other fine-grain factions from the concrete mix, thus increasing its resistance to washout.

6 Conclusions

The washout resistance of underwater concrete mixes is mostly shaped by chemical admixtures that increase viscosity. In most cases, these are synthetic and natural organic polymers soluble in water. Polymer chains intertwine cement grains and other fine particles present in cement paste, thereby leading to the increase in the paste's viscosity and reduction in of concrete mix washout.

In order to ensure greater washout resistance of underwater concrete mixtures, it is recommended to use anti-washout admixtures in combination with fine particles introduced with mineral additives with a high specific surface.

It was demonstrated that the washout resistance of an underwater concrete mix based on cement containing blast-furnace slag improves along with the increase in the slag's fine particles content (especially smaller than 20 μm) and with its increasing quantity in cement. It allows to reduce the concrete mix washout even by about 70%.

The washout resistance of an underwater concrete mixture can also be improved by introducing a small quantity of SiO_2 nanoparticles. The addition of 0.5% nanoparticles in relation to the cement's weight reduces the concrete mixes' washout by as much as 40%.

References

1. Yao SX, Gerwick BC (2004) Underwater concrete, construction issues. Concrete International, pp 60–64
2. Kańka S, Sołtysik R (2010) Examples of the use of underwater concreting technology in the renovation of hydrotechnical structures. Przegląd Budowlany 7–8:54–57
3. Kańka S, Rysiewicz M (2011) Experience with concreting the foundation of a cylindrical intake tower using underwater concrete. Inżynieria i Budownictwo 67(10):519–522
4. Okamura H, Ozawa K (1995) Mix design for self-compacting concrete. Concr Libr JSCE 25
5. Okamura H, Ozawa K, Ouchi M (1996) Self-compacting high performance concrete. Collcted Papers (University of Tokyo Department of Civil Engineering) 34:33–41
6. Khayat KH, Sonebi M (2001) Effect of mixture composition on washout resistance of highly flowable underwater concrete. ACI Mater J 289–295
7. Khayat KH, Assaad J (2003) Relationship between washout resistance and rheological properties of high-performance underwater concrete. ACI Mater J 185–193
8. Yao SX, Gerwick BC (2004) Underwater concrete, part 3: proper mixture proportioning. Concr Int 77–82
9. Sonebi M, Tamimi AK, Bartos PJM (2000) Application of factorial models to predict the effect antiwashout admixture, superplasticizer and cement on slump, flow time and washout resistance of underwater concrete. Mater Struct 33:317–323
10. Sam XY, Berner DE, Gerwick BC (1999) Assesment of underwater concrete technologies for in-the-wet construction of navigation structures. US Army Corps of Engineers, USA
11. Ramachandran VS (1984) Concrete admixtures handbook. Noyes Publications, Park Ridge, N.J., pp 528–533
12. Kawai T (1987) Non-Dispersible underwater concrete using polymers, marine concrete. In: International congress on polymers in concrete, Brighton, England
13. Khayat KH (1995) Effects of antiwashout admixtures on fresh concrete properties. ACI Mater J 92(2):161–174
14. Van Chanh N (2008) Design and construction of antiwashout underwater concrete. In: The 3rd ACF international conference ACF/VCA, Ho Chi Minh, Vietnam, pp 356–364
15. Khayat KH, Yahia A, Sonebi M (1988) Applications of statistical models for proportioning underwater concrete. In: 4th International conference on recent advanced in concrete technology, Japan, pp 95–113
16. Grzeszczyk S, Jurowski K (2015) Effect of metakaolin addition on the properties of self compacting concrete for underwater construction. Cement Wapno Beton 6:393–400
17. McLeish A (1994) Underwater concreting and repair. Halsted Press, New York, USA
18. Yousri KM (2008) Self-flowing underwater concrete mixtures. Mag Conc Res 60(1):1–10
19. Kojouri SJ, Abang Ali AA, Farzadnia N (2016) Effect of colloidal nano silica as a viscosity modifying agent in self compacting concrete with high volume POFA. Adv Mater Res 1133:491–495
20. Yao SX, Gerick BC, Underwater Concrete—mix design and construction particles, Inc. San Francisco, CA, USA
21. Grzeszczyk S, Jurowski K, Bosowska K, Grzymek M (2019) The role nanoparticles in decrease washout of underwater concrete. Constr Build Mater 203:670–678
22. Grzeszczyk S, Jurowski K, Kowalska J (2019) Effects of adding granulated blast-furnace slag to cement on the washout of concrete mixes in underwater construction. Cement Wapno Beton 6:515–524

The Influence of Grinding Method on the Particle Size Distribution of Selected Waste Materials Used in Concrete Technology

A. Kaleta-Jurowska, K. Jurowski, and J. Kowalska

Abstract In this paper, the properties of ground granulated blast furnace slag and fly ash, ground using a ball mill were examined. The influence of grinding parameters on the properties of the obtained material was observed by testing particle size distribution and Blaine specific surface. Rheological parameters of binders containing ground materials were also determined. The importance of binder fineness in concrete technology was discussed.

Keywords Concrete · Waste materials · Mineral additives

1 Introduction

1.1 *Utilization of Waste Materials in Concrete Production*

Production of concrete is a branch of industry, where waste materials coming from various industry types can be utilized in an effective way [1]. These materials, when properly prepared and mixed with the Portland cement clinker, are so-called mineral additives and constitute a perfect supplement to Portland cements commonly used for concrete production.

Mineral additives used as a substitute of the clinker, which react with calcium hydroxide produced during cement hydration, are often called "Supplementary Cementitious Materials" (SCMs). This group includes fly ash, granulated blast furnace slags (GBFS) and natural pozzolans, as well as strongly pozzolanic additives, such as silica fume, metakaolin, etc. [2]. These additives are classified in PN-EN 206 standard as type II additives.

There are many possibilities to use the waste materials from various industry types in concrete. Waste materials may be used as substitutes for aggregates (ground glass, some plastics, ceramics, rubber, agricultural waste, concrete waste). However, more

A. Kaleta-Jurowska (✉) · K. Jurowski · J. Kowalska
Faculty of Civil Engineering and Architecture, Opole University of Technology, Opole, Poland
e-mail: a.kaleta-jurowska@po.edu.pl

Z. Zembaty et al. (eds.), *Environmental Challenges in Civil Engineering*, Lecture Notes in Civil Engineering 122,
https://doi.org/10.1007/978-3-030-63879-5_11

often the waste materials are used as mineral additives to the cement (fly ash, silica fume, ground-granulated blast-furnace slag) [3].

New solutions are still being searched for aiming at utilization of various types of waste. The waste material of a great significance for the environment is tire rubber. As data presented in [4] show the approximate weight of scrap-tires generated annually only in the USA is estimated at 3.6 million tons. Tests showed that concretes with tire rubber addition had much lower density, better ductility, better resistance to impacts and better sound-proof properties. However, in case of concretes made of normal cements, a drastic reduction of the compressive strength is observed [5].

A natural substitute of aggregate, which allows to get light concrete, are coconut shells. There are also research studies conducted to replace sand and a part of cement with calcareous tuffs. Results of research often demonstrate a beneficial impact of individual waste materials on selected properties of the concrete made with their application, however, the environment-friendly aspect of waste materials utilization is emphasized [6].

To put it simply, production of the Portland cement includes the following phases: grinding of raw materials, their firing in temp. 1500 °C and then grinding of clinker with gypsum. It is estimated that energy needed to produce one ton of the cement is about 4000 MJ. As most of mineral additives requires grinding only, their use saves a lot of energy [7].

Production of Portland cement is one of the most important factors in CO_2 world-wide emission strategy [8]. Statistic data show that in 2017 China had the largest production of the cement equal to 2.38 Bnt. The next places are taken by India (423 Mt/yr), and then the USA (120.5 Mt/year) [9]. A high CO_2 emission is caused by the fact, that the Portland cement clinker production requires combustion of solid fuels (for calcination of limestone in high temperatures) [2]. Production of the clinker also requires transport of raw materials and electricity supply to the cement plant. Therefore it is estimated that the cement industry in total accounts for 8% of the global emission of carbon dioxide [10, 11]. Reduction of these values is possible due to utilization of mineral additives as substitutes for the clinker [12] or through application of waste materials such as phosphogypsum and waste limestone sludge, waste stone sludge, iron oxide sludge, and spent catalyst as raw materials in the production of the so-called eco-cement [13, 14].

1.2 Influence of Selected Mineral Additives on Concrete Properties

The use of mineral additives in concretes has a beneficial effect not only on reduction of CO_2 emission to air by substitution of the Portland cement clinker with industry by-products [12]. Another advantages of their use is higher durability of the concrete made, it's denser structure or better resistance to corrosion [12, 15, 16].

Results of tests show that the content of pozzolanic additives in concrete increases the concrete strength at a later curing period. It is an effect of C-S-H phase formation as a result of the pozzolanic reaction [12]. The use of mineral additives to cement also causes reduction of porosity of hardened cement pastes. Tests demonstrate that pastes with mineral additives had lower porosity than those made of the Portland cement. In addition, porosity is even lower in case of pozzolanic additives, where reduction of pore diameters is also observed [12].

Concretes, in particular special concretes (e.g. HPC, SCC) require a large quantity of a binder, which significantly raises costs of their production, and in consequence it also has a negative impact on the environment. As it is commonly known, concretes with the high content of the Portland cement are also more sensitive to shrinkage cracking. A solution for this problem is application of proper mineral additives, such as blast furnace slags, fly ash or fillers such as limestone powder or stone powder [17]. Mineral additives mentioned above have also a beneficial impact on consistency and workability of the concrete mix, process of cement hydration, and in consequence on properties of the hardened concrete [18].

According to classification given in PN-EN 196-1 standard, there are cements marked as CEM II, containing up to 35% of the mineral additive, CEM III (blast-furnace cement) with addition of ground granulated blast furnace slag in the amount from 35% up to even 95 wt% of cement. CEM IV cements contain mineral additives of pozzolanic properties, while CEM V cements may contain up to 60% of various mineral additives.

Silica fumes are a by-product during production of silicon metal, especially iron and silicon alloys in electric arc furnaces in very high temperatures (2500–2800 °C). Silica fumes are amorphous substance (approx. 85% of amorphous phase) consisting of very fine particles. The average size of particles is ca. 0.2 μm. The specific surface area of silica fume is larger by one order of magnitude than the specific surface area of the cement [19].

Due to the content of amorphous silica as the mineral additive to concrete, silica fumes show very strong pozzolanic properties [20]. The silica in reaction with the calcium hydroxide causes formation of C-S-H phase (responsible for the strength) and in consequence the gradual growth of the strength of the concrete made this way [20].

Unique properties of silica fume also have a significant impact on properties of the fresh concrete mix. They increase the cohesion which resist segregation of concrete ingredients. Silica fumes also cause limitation of so-called bleeding, i.e. water and fine fractions flowing out of the concrete mix [21–23].

A popular mineral additive used in the cement production is the fly ash. It is obtained as a by-product in coal fired power plants. Over 80% of unburned carbon particles leave the furnace together with flue gas. They are collected in result of dust separation by means of electrostatic or mechanical processes. In majority they have a spherical shape with dimensions not exceeding 200 μm [19, 20].

Addition of fly ashes increases flowability of the concrete mix without raising its production costs. It also allows to reduce hydration heat in the concrete built in [24]. Fly ash that raises the concrete strength contains significant quantities of silicon and aluminium oxides, which makes it the strongly pozzolanic additive after a longer curing period. In consequence the concrete with this additives has better durability [25].

Beneficial effects can be also achieved with various mineral additives applied simultaneously. Tests presented in the paper [26] demonstrated that concretes made with the use of fly ash and silica fume simultaneously reached the strength higher by even 40%, compared to concretes with fly ash only. Tests show that application of these additives allows to improve workability of the concrete. Furthermore, authors of [27] proved that it was more justified economically than control of concrete consistency with the use of a superplasticizer.

Another waste material from industry is the fly ash coming from combustion in fluidized bed boilers. However, the use of this material, is problematic, due to a high content of calcium sulphate, which may cause delayed production of ettringite, and in consequence destruction of the concrete. They are also heterogeneous, as their chemical composition varies depending on a place of combustion and an origin of the material being combusted [28].

Steel slag is an industrial waste from either the conversion of iron to steel in a basic oxygen furnace or from melting scrap to make steel in an electric arc furnace. Its composition is made of oxides such as CaO, MgO, SiO_2, and FeO, which makes it have binding properties and it is ideal for the use as the mineral additive to the Portland cement. It shall be noted that the CO_2 emission during its production is much lower than during production of the Portland cement clinker [29].

A very important feature of concrete made of cement with addition of the ground granulated blast furnace slag is a fact that it emits much less heat than the concrete made of the Portland cement [7]. Concrete made with the use of blast furnace slags has a slower strength growth rate (compare to Portland cement based concrete), but its final strength is higher. It also shows better durability regardless of its exposure. It is due the fact that a microstructure of hardened binder is compacted and sealed. It was also found that the use of the blast furnace slag as the mineral additive to cement is more effective than its use as the aggregate, e.g. for subsoil stabilization [7].

It is estimated that the energy needed to grind the granulated blast furnace slag is ten times lower than the energy needed for production of the Portland cement clinker [7]. Due to the fact that production of the granulated blast furnace slag requires less power consumption, concrete with this additive can be deemed as more environment friendly [29].

Addition of granulated blast furnace slag to cement has also a beneficial effect on workability of self-compacting concrete mixes. Tests presented in the paper [30] show that the optimum quantity of slag to the cement mass is 20%. However, it should be remembered, that concretes with this additive reach lower early strength. Ground granulated blast furnace slag has a good influence on later strength, which was demonstrated by tests after 56 and 90 days of curing.

This paper analyses the impact of grinding (fineness) of various types of mineral additives: fly ash and ground granulated blast furnace slag on properties of mortars. Physical properties of additives and impact on properties of fresh and hardened

Table 1 Chemical composition of cement CEM I 42.5 R, slag and fly ash

Components	SiO_2	Al_2O_3	Fe_2O_3	CaO	MgO	K_2O	Na_2O_e	SO_3
CEM I 42.5 R (CI)	19.5	6.2	2.8	62.8	1.4	0.8	0.7	2.6
Ground granulated blast furnace slag (S)	38.6	7.8	1.0	40.0	9.1	0.3	0.4	0.1
Fly ash (V)	40.2	6.0	2.0	43.2	4.7	–	–	0.1

mortars were tested. The impact of mineral additives fineness on mortar compressive strength was also analysed.

2 Materials

Materials used for testing: cement—CEM I 42.5 R (CI) and mineral additives: fly ash (V), and ground granulated blast furnace slag (S). Chemical composition of binders tested is presented in Table 1.

Binders used for testing of the compressive strength obtained from mixing of cement CEM I 42.5 R (CI) and fly ash (V) or ground granulated blast furnace slag (S) in amount of 30% by mass. Samples of binders were homogenized for approx. one hour.

In order to compare the impact of mineral additives of various fineness, fly ash and granulated blast furnace slag were additionally ground in a ball mill. 50 balls were used for grinding—diameter 25 mm and weight ca. 66 g each, and the grinding process was continued for 4 h.

Particle size distribution of materials used for testing is presented in Fig. 1. The graphs present a differential curve and a cumulative curve of particle size distribution.

The specific surface area was also determined for the binders tested with application of the Blaine method acc. to PN-EN 196-6:2011. Obtained values of specific surface area of the materials used are presented in Table 2. Differences in specific surface area values are very significant, which also indicates a large difference in fineness of individual materials.

Six different mixes of binder in total were prepared for testing. Composition of mortars tested is presented in Table 3. Mortars made of the binder that was additionally ground in the lab were marked with "G" symbol.

3 Testing Methods

Cement particle size distribution was tested by means of a particle size laser analyzer—Mastersizer 3000 with a wet dispersion method. Isopropyl alcohol was

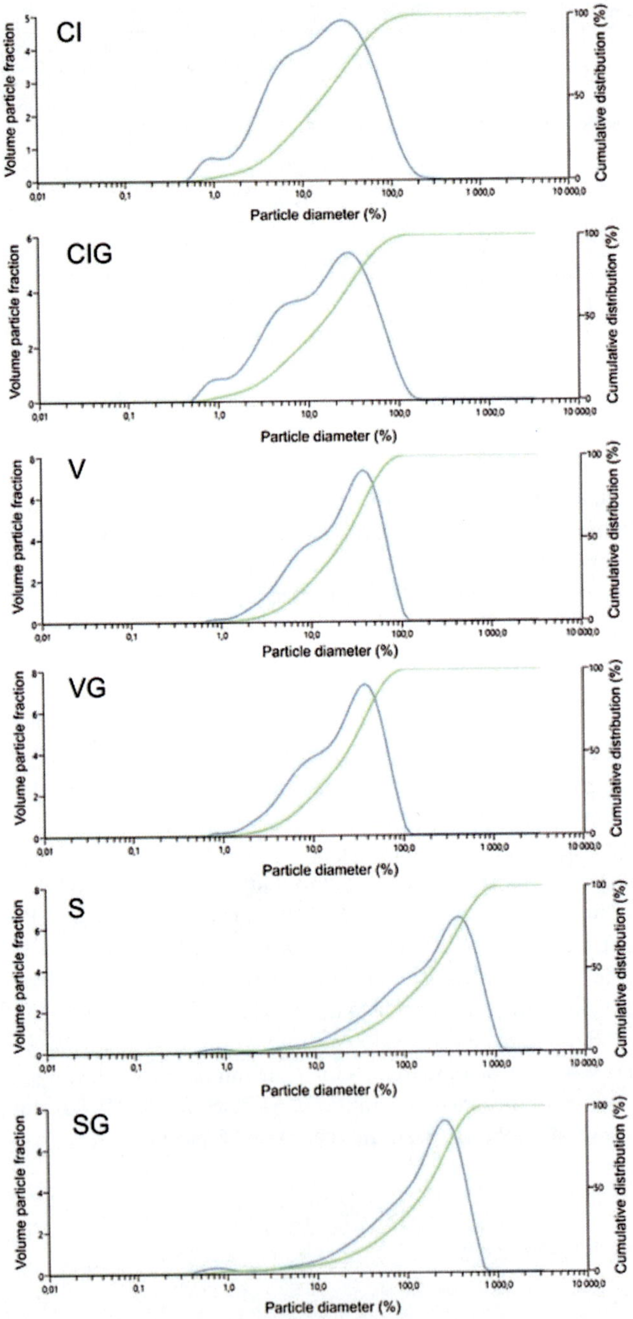

Fig. 1 Particle size distribution of non-ground cement (CI), fly ash (V) and ground granulated blast furnace slag (S), as well as ground ones marked as CIG, VG and SG respectively

Table 2 Test results of density and specific surface area acc. to Blaine method

Designation	Specific density (g/cm^3)	Specific surface area (cm^2/g)
CI	3.10	4650
CIG	3.10	5480
V	2.30	2380
VG	2.36	2640
S	2.83	330
SG	2.86	540

Table 3 Determination and composition of mortars tested (g)

Designation	Cement CI	Grinded cement CIG	Fly ash V	Grinded fly ash VG	Ground granulated blast furnace slag S	Grinded ground granulated blast furnace slag SG
CI	450	–	–	–	–	–
CIG	–	450	–	–	–	–
CI + 30%V	315	–	135	–	–	–
CI + 30%VG	315	–	–	135	–	–
CI + 30%S	315	–	–	–	135	–
CI + 30%SG	315	–	–	–	–	135

used as a dispersant. Volumetric particle size distribution was determined by means of the differential curve, as well as cumulative curve of grain-size distribution.

Consistency of mortars was tested in line with PN-85/B-04500 standard. This method consists in immersion of a slump cone under its own weight in the mortar placed in a suitable cone-shaped container. A result of this test is the cone immersion depth. Consistency was tested directly after mixing of ingredients.

The flexural and compressive strength of mortars were tested in accordance with PN-EN 196-1 standard, on beam samples with dimensions 160 × 40 × 40 mm. The flexural strength was determined during a three-point bending test at support spacing 100 mm. The compressive strength was determined on both ends formed after breaking. Testing was conducted after 28 days of mortars curing in water. Test results given are average values of tests conducted on three sample beams.

The measurement of the rheological parameters of the cement pastes were carried out using a rotational viscometer with co-axial cylinders, Haake Mars III type. The same procedure for the preparation samples for measurements and the same measurement conditions were maintained. During preparation and when conducting tests the temperature in the laboratory was kept at 20 ± 2 °C. After hand-stirring the mixing cement with water (3 min), the sample was immediately loaded into the cylinder for testing. The shear rate was increased and decreased in a range from 0 to 150 s^{-1} in 6 min. The results of the tests were presented as a flow curves. The measurements were performed at temperatures of 20.0 ± 0.1 °C, after 10, 30 and 60 min. Yield stress and plastic viscosity was calculated using the Bingham model. The area of the hysteresis loop of flow curves was calculated using a numerical integral method (trapezoidal rule).

4 Results of Testing and Their Analysis

Selected values from the particle size distribution testing with a laser method are presented in Table 4. A percentage of cement particles with a diameter below 10 μm, 50 μm and 100 μm and values Dv (10), Dv (50) and Dv (90) that stand for the particle diameter below which there are 10%, 50% and 90% of particles of material tested respectively.

Data presented in Table 4 show that 50% of particles have the diameter below 17.1 μm, whereas after additional grinding this diameter is 15.9 μm. Analogically, for the fly ash a half of particles is smaller than 33.1 μm before grinding and 23.8 μm after grinding.

The laser analysis of grain-size distribution demonstrated that the tested blast furnace slag had much larger particles. It is clearly visible in graphs of grain-size distribution presented in Fig. 1, where the cumulative size distribution curve is shifted towards larger particles compared to cement and fly ash. That phenomenon is also confirmed by numerical data contained in Table 4, which say that particles of granulated blast furnace slag are almost 10-times larger.

Comparison of data in Table 4 also allows to state, how the grinding in the ball mill affects the particle size distribution of materials tested. Fly ash seems to be the most prone to grinding, and its diameter Dv(90) after the grinding reduced by over a half (Fig. 2). Diameters of blast furnace slag particles were also clearly reduced,

Table 4 Analysis of particle size distribution of binders tested

Material	Dv(10)	Dv(50)	Dv(90)	Content of particles		
				<10 μm	<50 μm	<100 μm
	μm			(%)		
Cement CI	3.1	17.1	67.5	33.4	82.6	96.5
Grinded cement CIG	2.7	15.9	55.8	37.7	87.3	98.9
Fly ash V	6.1	33.1	146	22.3	60.2	79.9
Grinded fly ash VG	5.1	23.8	59.4	24.4	83.3	99.8
Ground granulated blast furnace slag S	27.4	210	598	3.7	17.4	30.8
Grinded ground granulated blast furnace slag SG	22.4	161	399	5.0	20.5	35.4

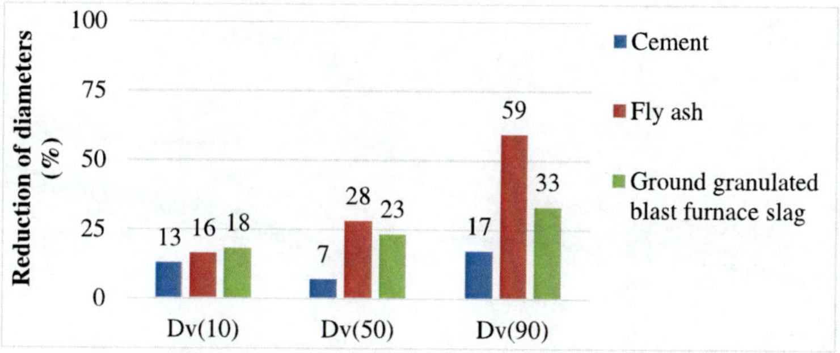

Fig. 2 Reduction of diameters Dv(10), Dv(50) and Dv(90) for CEM I, fly ash and granulated blast furnace slag, due to grinding in the ball mill, expressed in per cents

however, they were incomparably larger than particles of the fly ash and cement CEM I 42.5 R.

Results of the specific surface area tests with the Blaine method are presented in Table 2. This surface area is definitely the largest in case of cement CEM I 42.5 R, it gets an intermediate value for fly ash, and it is relatively low for the granulated blast furnace slag. Values of specific surface area also reflect the material fineness, which can be seen by comparing values from Tables 2, 3 and 4.

Based on measurement results obtained, flow curves of cement pastes were made (Fig. 3). Rheological parameters (yield stress and plastic viscosity) of cement pastes tested were determined by fitting the Bingham model to the descending flow curve. A clear increase of rheological parameters was found as particles of the finer binder was used. The highest values of plastic viscosity were observed in case of pastes with addition of fly ash (Table 5). Whereas the highest flowability was demonstrated by pastes with the ground granulated blast furnace slag. Obtained values of rheological parameters correlate with values of the specific surface area of mineral additives applied, which is understandable, as the water/binder ratio used for all testing pastes was 0.4.

The analysis of the value of hysteresis loop surface area allows to state that the highest values were obtained in case of pastes with addition of fly ash (Table 5). It proves that this additive has a strong influence on thixotropic behaviour of cement paste.

Results of flexural and compressive strength testing of mortars are presented in Table 6. In order to keep comparability of strength test results, the water/binder ratio maintained in all mortars was 0.5. Table 6 also presents a result of mortars consistency testing.

Consistency of mortars was different depending on the fineness and the quantity of the additive to cement. While keeping the constant water/binder ratio, the highest flowability was observed in case of mortars with addition of fly ash. This can be explained by a spherical shape of fly ash particles, which allows to get better flowability compared to the Portland cement and granulated blast furnace slag.

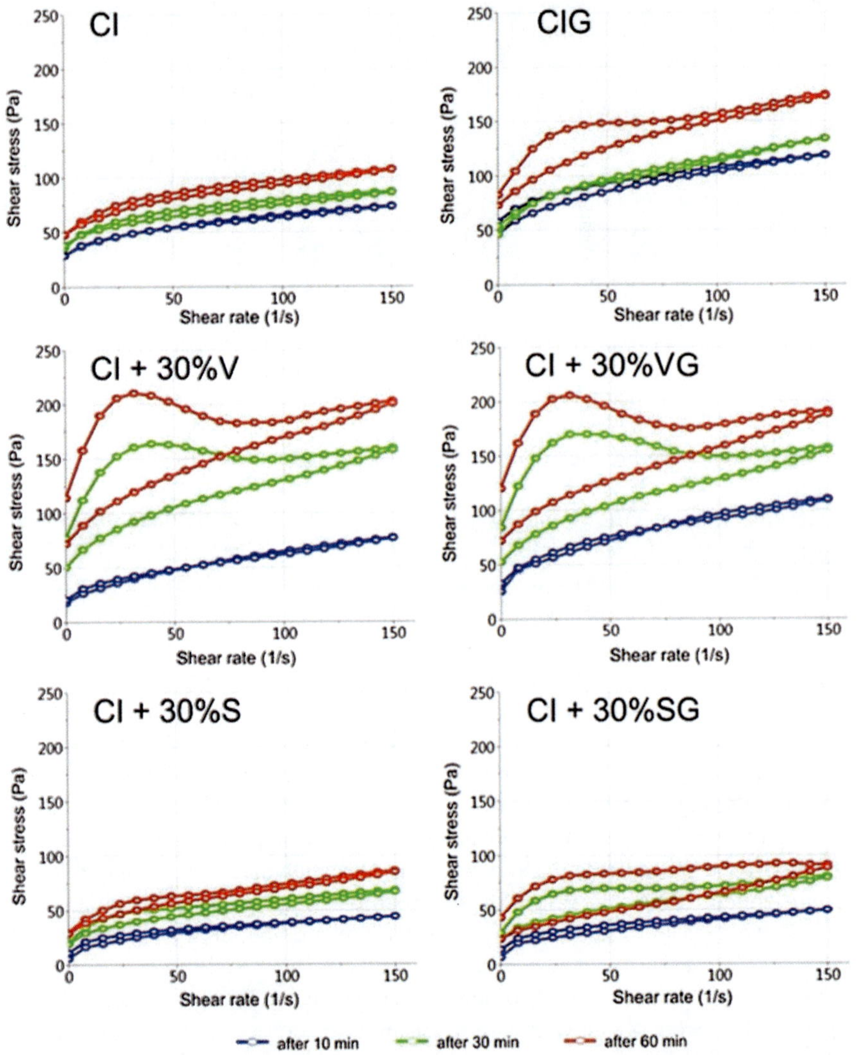

Fig. 3 Flow curves of cement pastes made of cement (CI), ground cement (CIG), and cement (CI) with addition of 30% of fly ash non-ground (V) or ground (VG) and cement (CI) with addition of 30% of ground granulated blast furnace slag (S) or non-ground and ground (SG)

Comparison of flexural and compressive strength values of mortars after 28 days, depending on the specific surface area of the cement or the mineral additive used, is presented in Figs. 4 and 5. In case of the compressive strength, the relationship between the fineness degree (determined by a value of the specific surface area acc. to Blaine) and the strength achieved is clearly visible.

Table 5 Results of rheological testing

Sample	Testing after time	Rheological parameters		Surface area of hysteresis loop (Pa/s)
		Yield stress (Pa) plastic viscosity	(Pa·s)	
CI	10	44.6	0.21	158
	30	52.2	0.23	570
	60	65.2	0.29	640
CIG	10	66.2	0.37	880
	30	75.5	0.40	–
	60	98.2	0.52	2100
CI + 30%V	10	33.8	0.28	31
	30	43.4	0.34	3300
	60	54.4	0.38	3740
CI + 30%VG	10	57.0	0.35	–
	30	78.5	0.51	5890
	60	95.4	0.62	6590
CI + 30%S	10	25.4	0.13	–
	30	33.1	0.22	855
	60	42.3	0.28	690
CI + 30%SG	10	27.9	0.15	–
	30	37.1	0.27	1870
	60	39.0	0.37	3986

Table 6 Results of mortars consistency testing, compressive and flexural strength testing

Sample	Consistency (cm)	Strength	
		Flexural (MPa)	Compressive (MPa)
CI	4.0	6.54	45.4
CIG	5.0	7.15	55.2
CI + 30%V	6.5	5.87	32.6
CI + 30%VG	6.5	5.55	35.2
CI + 30%S	3.5	4.86	25.1
CI + 30%SG	3.0	4.66	26.5

While interpreting the data shown in Figs. 4 and 5 it shall be mentioned that additives of fly ash and ground granulated blast furnace slag were added in 30% by mass, and the specific surface area presented in graphs refers to the additive itself. Yet, the increase of the compressive strength of mortars can be observed along with the increase of the specific surface area of the mineral additive used. Tests did not prove a similar tendency in the flexural strength of mortars, but differences in

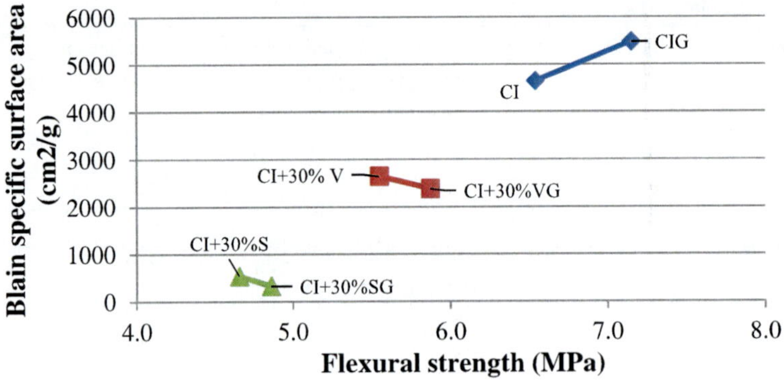

Fig. 4 Flexural strength of mortars depending on the specific surface area of cements and mineral additives applied

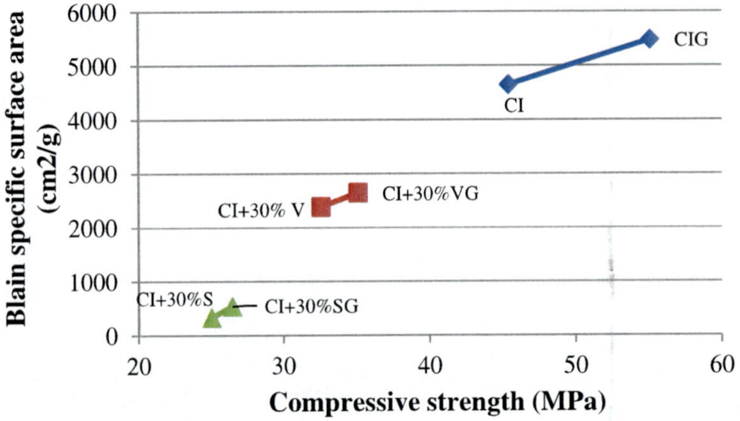

Fig. 5 Compressive strength of mortars depending on the specific surface area of cements and mineral additives applied

flexural strength values obtained in case of ground and non-ground mortars are not significant.

5 Summary

Waste materials from the industry, such as fly ash and granulated blast furnace slag are suitable for use in the concrete technology. Tests of mortars demonstrated that replacement of 30% of the binder with the fly ash and ground granulated blast furnace slag results in reduction of the compressive strength after 28 days by 28% and 45%

respectively in case of non-ground additives, and 22% and 40% respectively in case of additionally ground additives.

Reduction of the compressive strength due to replacement of a part of cement with mineral additives was observed, however, does not disqualify their use in the concrete technology. Since it contributes to reduction of the carbon dioxide emission to air, while keeping still satisfactory strength parameters. In addition, it shall be expected that after a longer curing time, differences in strength values of mortars obtained would be reduced because, as it is commonly known, these additives reduce the strength growth rate in the early period of curing.

Results of compressive strength tests indicate that the increase of cement fineness affects the increase of this parameter (in this case by ca. 22%). A similar impact was observed when fly ash and blast furnace slag was added, for which the grinding process increased the compressive strength obtained by 8% and 6% respectively.

Rheological tests demonstrated that along with the increase of mineral additives fineness, the increase of rheological parameter values was observed in case of all pastes tested. Also a significant growth of the hysteresis loop surface area was observed in case of the fly ash additive.

References

1. Tavakoli D, Hashempour M, Heidari A (2018) Use of waste materials in concrete: a review. Pertanika J Sci Technol 26:499–522
2. Damtoft JS, Lukasik J, Herfort D, Sorrentino D, Gartner EM (2008) Sustainable development and climate change initiatives. Cem Concr Res 38:115–127. https://doi.org/10.1016/J.CEMCONRES.2007.09.008
3. Agarwal SK (2006) Pozzolanic activity of various siliceous materials. Cem Concr Res 36:1735–1739. https://doi.org/10.1016/J.CEMCONREoS.2004.06.025
4. Siddique R, Naik TR (2004) Properties of concrete containing scrap-tire rubber—an overview. Waste Manag 24:563–569. https://doi.org/10.1016/J.WASMAN.2004.01.006
5. Misra AK, Mathur R (2007) Magnesium oxychloride cement concrete. Bull Mater Sci 30:239–246. https://doi.org/10.1007/s12034-007-0043-4
6. Hadj Aissa Y, Goual I, Benabed B (2020) Mix-design and properties of self-compacting concrete made with calcareous tuff. J Build Eng 27:100997. https://doi.org/10.1016/J.JOBE.2019.100997
7. Shi C, Qian J (2000) High performance cementing materials from industrial slags—a review. Resour Conserv Recycl 29:195–207. https://doi.org/10.1016/S0921-3449(99)00060-9
8. Moya J, Pardo N, Mercier A (2010) Energy efficiency and CO_2 emissions: prospective scenarios for the cement industry. https://doi.org/10.2790/25732
9. Edwards P (2017) Global cement top 100 report 2017–2018. Glob Cem, Mag
10. Olivier J, Peters J, Janssens-Maenhout G (2012) Trends in global CO_2 emissions 2012 report. https://doi.org/10.2788/33777
11. Bakhtyar B, Kacemi T, Nawaz MA (2017) A review on carbon emissions in Malaysian cement industry. Int. J Energy Econ Policy 7:282–286
12. Senhadji Y, Escadeillas G, Mouli M, Khelafi H (2014) Benosman, Influence of natural pozzolan, silica fume and limestone fine on strength, acid resistance and microstructure of mortar. Powder Technol 254:314–323. https://doi.org/10.1016/J.POWTEC.2014.01.046

13. Lin K-L, Lo K-W, Hung M-J, Cheng T-W, Chang Y-M (2017) Recycling of spent catalyst and waste sludge from industry to substitute raw materials in the preparation of Portland cement clinker. Sustain Environ Res 27:251–257. https://doi.org/10.1016/J.SERJ.2017.05.001
14. Islam GMS, Chowdhury FH, Raihan MT, Amit SKS, Islam MR (2017) Effect of phosphogypsum on the properties of Portland cement. Procedia Eng 171:744–751. https://doi.org/10.1016/J.PROENG.2017.01.440
15. Bertron A, Duchesne J, Escadeillas G (2005) Accelerated tests of hardened cement pastes alteration by organic acids: analysis of the pH effect. Cem Concr Res 35:155–166. https://doi.org/10.1016/J.CEMCONRES.2004.09.009
16. Oueslati O, Duchesne J (2012) The effect of SCMs and curing time on resistance of mortars subjected to organic acids. Cem Concr Res 42:205–214. https://doi.org/10.1016/J.CEMCONRES.2011.09.017
17. Sonebi M (2004) Medium strength self-compacting concrete containing fly ash: modelling using factorial experimental plans. Cem Concr Res 34:1199–1208. https://doi.org/10.1016/J.CEMCONRES.2003.12.022
18. Huang W, Kazemi-Kamyab H, Sun W, Scrivener K (2017) Effect of cement substitution by limestone on the hydration and microstructural development of ultra-high performance concrete (UHPC). Cem Concr Compos 77:86–101. https://doi.org/10.1016/J.CEMCONCOMP.2016.12.009
19. Grzeszczyk S, Janowska-Renkas E, Konopka E, Marynowicz A, Matuszek-Chmurowska A, Mordak A, Skaliński B (2012) Materiały Budowlane, Politechnika Opolska, Wydział Budownictwa, Opole
20. Mardani-Aghabaglou A, İnan Sezer G, Ramyar K (2014) Comparison of fly ash, silica fume and metakaolin from mechanical properties and durability performance of mortar mixtures view point. Constr Build Mater 70:17–25. https://doi.org/10.1016/J.CONBUILDMAT.2014.07.089
21. Poon CS, Kou SC, Lam L (2006) Compressive strength, chloride diffusivity and pore structure of high performance metakaolin and silica fume concrete. Constr Build Mater 20:858–865. https://doi.org/10.1016/J.CONBUILDMAT.2005.07.001
22. Elahi A, Basheer PAM, Nanukuttan SV, Khan QUZ (2010) Mechanical and durability properties of high performance concretes containing supplementary cementitious materials. Constr Build Mater 24:292–299. https://doi.org/10.1016/J.CONBUILDMAT.2009.08.045
23. Giner VT, Ivorra S, Baeza FJ, Zornoza E, Ferrer B (2011) Silica fume admixture effect on the dynamic properties of concrete. Constr Build Mater 25:3272–3277. https://doi.org/10.1016/J.CONBUILDMAT.2011.03.014
24. Siddique R (2011) Properties of self-compacting concrete containing class F fly ash. Mater Des 32:1501–1507. https://doi.org/10.1016/J.MATDES.2010.08.043
25. Yazıcı H, Aydın S, Yiğiter H, Baradan B (2005) Effect of steam curing on class C high-volume fly ash concrete mixtures. Cem Concr Res 35:1122–1127. https://doi.org/10.1016/J.CEMCONRES.2004.08.011
26. Nochaiya T, Wongkeo W, Chaipanich A (2010) Utilization of fly ash with silica fume and properties of Portland cement–fly ash–silica fume concrete. Fuel 89:768–774. https://doi.org/10.1016/J.FUEL.2009.10.003
27. Yahia A, Tanimura M, Shimabukuro A, Shimoyama Y (1999) Effect of rheological parameters on self compactability of concrete containing various mineral admixtures. In: International symposium on 1st, self-compacting concrete, Stockholm, pp 523–536
28. Janowska-Renkas E, Kowalska J (2018) Use of fly ash from fluidized bed boilers in clinker-slag-ash based binders. MATEC Web Conf 174:1–11. https://doi.org/10.1051/matecconf/201817402002
29. Patra RK, Mukharjee BB (2016) Fresh and hardened properties of concrete incorporating ground granulated blast furnace slag—a review. Adv Concr Constr 4:283–303. https://doi.org/10.12989/acc.2016.4.4.283
30. Boukendakdji O, Kadri E-H, Kenai S (2012) Effects of granulated blast furnace slag and superplasticizer type on the fresh properties and compressive strength of self-compacting concrete. Cem Concr Compos 34:583–590. https://doi.org/10.1016/J.CEMCONCOMP.2011.08.013

Contemporary Churches of Czestochowa—Architectural and Urbanistic Aspects

A. Repelewicz

Abstract Sacral architecture is a significant urban element in every city. For centuries church buildings have been one of the most important public facilities. In Czestochowa too, they have had a huge impact on the arrangement of public spaces. The most important religious building in Czestochowa is Jasna Góra monastery. The oldest parish church, St. Sigismund's, is located exactly opposite the monastery along the central axis of the city. For several centuries, Czestochowa had only one parish, covering the city and the surrounding area. Only two churches were erected in the town before the nineteenth century. Further Catholic sacral buildings were built in the nineteenth century, when Czestochowa was intensively developing. After World War II, there was a stagnation in the field of religious architecture and by 1989 only a few further churches had been built. After 1989, there were no longer obstacles to applying for building permits for sacral buildings. Since that time, 17 new church buildings have been erected and finished in Czestochowa. This paper will present the issue of the spatial location of churches in the city's districts. The size and architectural style of these objects will also be discussed.

Keywords Sacral architecture · Urbanistic

1 Materials and Methods

The analyses are based on the authors own researches conducted among 2010–2018 in the Archdiocese of Czestochowa. A fragment of wider research is presented in the paper.

Churches drafts, information about estimate year outlays and data from parish records were collected from each parish and considered. Interviews with priests were conducted due to obtain information about usefulness of each part of construction. The building designs, if preserved, were examined. If these were not preserved, building inventory was taken. The total area of sacral buildings, their architectural

A. Repelewicz (✉)
EDP Sciences, Czestochowa University of Technology, Czestochowa, Poland
e-mail: aleksandra.repelewicz@pcz.pl

Z. Zembaty et al. (eds.), *Environmental Challenges in Civil Engineering*, Lecture Notes in Civil Engineering 122,
https://doi.org/10.1007/978-3-030-63879-5_12

style, functionality, type of construction and a lot of other data were collected. For the purposes of this paper the distances between new church buildings were analyzed over the years, based on the GPS coordinates of each structure.

2 The Oldest Churches of Czestochowa

St. Sigmund's is the oldest Catholic parish in Czestochowa. The first historical mention of the parish of St. Sigmund's appears in a list of saints from 1325–27. This historic church, in use today in the old town of Czestochowa, was erected after 1356, after city rights were granted. For several centuries, Czestochowa had only one parish, covering the city and the surrounding area. There was also a church, located south of Jasna Góra, on the former road from the monastery to the village of Kawodrza. This was St. Barbara and St. Andrew's [erection 1637, benediction 1642, consecration 1643], which only became a parish church in 1891. Additionally, one of the oldest sacral buildings in the city is the church of St. Roch and Sebastian [erection 1641, benediction 1642, consecration 1680]. It is a small church located in a cemetery.

Further Catholic sacral buildings were built in the nineteenth century, when Czestochowa was intensively developing. These were erected in the following order:

- Church of the Holy Name of Mary [erection 1859, benediction 1862];
- Church of St. James the Apostle, former Orthodox church of St. Cyril and Methodius [erection 1870, benediction 1872];
- Church of the Resurrection in the Mystery of Emmaus [erection and benediction 1898], erected as a cemetery chapel of Transfiguration of the Lord;
- Cathedral church of Holy Family [erection 1902, benediction 1907, consecration 1950]
- The filial church of the Dying Jesus [erection 1905, benediction in 1906], after extension in 1925 used as a parish (until 1954). [1, 2]

Almost all the above listed churches are in what is the centre of the city today. The Church of St. Sigmund is situated directly opposite Jasna Góra monastery, at the end of the main street of the city: The Avenue of the Blessed Virgin Mary. Two of the historic churches: Saint James' and the Holy Name of Mary's are located on the same street, which is the axis of the city, between the monastery and the church of St. Sigmund. The Archcathedral was erected in the vicinity of the oldest parish of the city, the churches of St. Barbara, St. Roch and the Dying Jesus are in the neighboring suburb of Podjasnogórska, and the church of the Lord's Resurrection is in the Tysiąclecie district, farthest from the city center. These were located away from the then urban buildings as was the cemetery on which it was located.

3 Churches in the Interwar Years

There were 8 catholic churches in Czestochowa (and Jasna Góra monastery) when Poland regained its independence in 1918. Their number had increased to eleven by the beginning of the Second War. Of these three, two were new objects and one was a converted fire station. They were:

- Church of St. Joseph the Craftsman, erected in the village of Raków transformed into an industrial settlement due to the construction of an ironworks [erection 1925, benediction 1933];
- Church of St. Anthony of Padua, built in the district of Ostatni Grosz, near textile factories [erection 1938, benediction and consecration 1956];
- Church of St. Andrew Bobola was adapted in 1940 from a fire station. The development and modernization of the facility took place in the second half of the 1940s.

These three churches are located outside the city center. The first two were built for residents of the new districts: Raków and Ostatni Grosz. Both served the faithful of these industrial districts, created due to the development of the metallurgy and textile industry in Czestochowa. The third one was located in the village of Kiedrzyn, now a city district.

4 Years 1945-2007

A further four Catholic churches were built in Czestochowa after the Second World War, in the 1950s. In the first decade after the war, this type of construction was not subject to restrictions and single building permits were granted. The first four objects in Table 1 are the so-called pre-Conciliar churches. The shape of their interior refers to the tradition from before the Second Vatican Council (Figs. 1 and 2).

Fig. 1 Church of St. Teresa of the Child Jesus

Fig. 2 Church of Raising the Holy Cross

After a period of relative freedom in obtaining permits for the construction of new churches, in the 1960s not even one building permit was issued in Czestochowa. At that time, the diocesan curia undertook a number of attempts to obtain permits for the construction of new churches, as evidenced by intensive correspondence with the voivodship and municipal authorities [2]. When the political situation changed after the events of December 1970, relations between church and state authorities improved. Permits were initially limited to the renovation of several churches in the Czestochowa diocese. It was not until 1973 that the first building permit was given for a church in the Trzech Wieszczów district. The next three permits were obtained after a further five years, in 1978 [3] (Figs. 3 and 4).

All four objects (no. 5–8 in Table 1) are post-Conciliar churches. They were designed in a modernist style, as very large, two-level buildings. One of the many adverse effects of the restrictions on obtaining planning permission for churches was that when building permission was given after years of failure to receive it, the bishop would decide to build a huge object, because the prospects of receiving further permission was uncertain. Today, after territorial changes and a split into smaller parishes, these objects are too large and create numerous operational problems.

No further building permits were issued until after 1980, when the political situation had changed. Trying to make up for the freeze on permits in the years 1980–1989, applications for the construction of many new facilities were submitted. As many as 15 new churches were built, doubling the total number of existing Catholic church facilities in the city at that time. Modernism was the dominant architectural style. Though two churches went against this trend: St. Stanislaus Kostka (post-modernism) and Saint. Elisabeth Hungarian (regionalism). A great number of the churches designed in these years were still very large objects (e.g. the church of St. Stanislaus, the church of the Transfiguration) or large objects (e.g. the church of Saint Apostles Peter and Paul and of Saint Francis of Assisi – Figs. 5 and 6). Gradually,

Fig. 3 Church of St. Adalbert

Fig. 4 Church of St. Stanislaus

however, the buildings became much smaller, the size of which was also adapted to meet the current needs of the parish (e.g. the church of St. Stanislaus Kostka).

After 1989, there were no more obstacles to applying for building permission for religious buildings. At that time, 17 new church buildings were erected and finished in Czestochowa. These are much smaller objects than those built in the time of

Table 1 Churches of Czestochowa 1945–2007

No.	The name of the church	Address	Year of the project creation	Designer	Years of the church construction
1	Divine Providence	ul. Warszawska 288	Before 1947	E. Usakiewicz	1947–1950
2	Raising the Holy Cross	Rynek Wieluński 29	Before 1949	ing. Sachse	1949–1954
3	St. Teresa of the Child Jesus	ul. Gościnna 2	1957	W. Kobzdej and J. Bajda	1957–1959
4	Virgin Mary Victorious	ul. Słowackiego 19/21	1975	A. Mazur	1975–1985
5	St. Adalbert Bishop and the Martyr	ul. Brzeźnicka 59	1978	A. Mazur	1978–1985
6	Sacred Heart of Jesus	ul. Kościelna 53/63	1978	A. Mazur	1978–1990
7	Virgin Mary Queen of Poland	ul. Konwaliowa 59	1980	A. and Z. Białkiewicz	1980–1984
8	St. Peter and Paul Apostles	ul. Wręczycka 128/134	1981	A. and Z. Białkiewicz	1981–1987
9	Transfiguration of the Lord	ul. Drzewna 40	Before 1981	No data	1981–1990
10	St. Albert Chmielowski	ul. Narcyzowa	1981	Aleksander Holas	1981–2001
11	St. Francis of Assisi	ul. Luba 26	1982	A. and Z. Białkiewicz	1982–1987
12	St. Stanislaus Bishop and the Martyr	ul. Boh. Katynia 40/42	1981	Jerzy Kopyciak	1983–2000
13	Visitation of the Virgin Mary	ul. Okólna 113	1983	B. Jezierski	1984–1990
14	St. Judah Tadeusz Apostle	ul. Kaczorowska 56	No data	A. and Z. Białkiewicz	1984–1987
15	Ascension of the Lord	ul. Sternicza 1	1985	Wojciech Rados	1986–2001
16	Bl. Urszula Ledóchowska	ul. Sportowa 80	1987	Jerzy Kopyciak	1987–1999
17	St. Stanislaus Kostka	ul. Gen. Dąbkowskiego 42	1987	Jerzy Kopyciak	1987–1988
18	St. Jadwiga the Queen	ul. Gen. Zajączka 6	1988	M. Gierlotka	1988–1991

(continued)

Table 1 (continued)

No.	The name of the church	Address	Year of the project creation	Designer	Years of the church construction
19	Immaculate Conception of the Blessed Virgin Mary	ul. Wielkoborska 123	1988	Adolf Penczor	1989–2010
20	Virgin Mary Helpers of the Faithful	ul. Hutników 68	1989	Z. Białkiewicz, B. Brągiel	1989–1997
21	St. Jadwiga Śląska	ul. Wypalanki 24	No project	brak danych	1991–1993
22	The First Martyrs of Poland	ul. Obr. Westerplatte 37	1991	Włodzimierz Domagała	1991–1998
23	Virgin Mary from the Carmel Mount	ul. Połaniecka 50A	1990	Zygmunt Fagas	1991–2000
24	St. Maksymilian Maria Kolbe	ul. Starzyńskiego 11	1990	Aleksander Holas	1992– 2002
25	St. Elizabeth Hungarian	ul. Tetmajera 3	1992	Ryszard Jurkowski	1992–2000
26	St. John the Baptiste	ul. Długa 78	1997	Janina Dunal	1992–1996
27	St. Jan Kantego	ul. Br. Czecha 2	1997	Maria Musialska	1993–2000
28	Conversions of Saint. Paul the Apostle	ul. Trzcinowa 33	1995	Maria Muskalska	1995–2000
29	Sacred Body and Blood of Christ	al. Jana Pawła II 82	1995	Włodzimierz Domagała	1995–2000
30	St. Jack	ul. Warszawska 452 A	About 1998	Stanisław Kwaśniewicz	1995–2000
31	St. Andrzej Świerad and Benedykt Pustelnik	ul. Mickiewicza 5/17	1996	Janina Dunal	1996–2000
32	Virgin Mary of Czestochowa	ul. Michałowskiego 28a	1995	Irena Czerwińska	1997–2000
33	St. Florian the Martyr	ul. Limanowskiego 120	1998	Z. and W. Domagała	1998–2000

(continued)

Table 1 (continued)

No.	The name of the church	Address	Year of the project creation	Designer	Years of the church construction
34	Blessed Rafał Kalinowski	ul. Żródlana 4	About 1998	Ziemowit Domagała	1999–2000
35	St. John of Dukla	ul. Mirowska 51/59	1994	Beata Król-Stroba	1999–2010
36	St. Kazimierz the Prince	ul. Brzeźnicka 45	1999	Maria Musialska	1999–2005
37	St. John Sarkander	ul. Cegielniana 7/11	2002	Ziemowit Domagała	2002–2007

Fig. 5 Church of St. Peter and Paul Apostles

the communism, and therefore better adapted to a dense parish network (currently there are 56 parishes in the city, in some of them the churches have not yet been completed) [1]. The stylistics of the newest churches are much more diverse. Apart from modernist buildings (e.g. churches: Saint Benedict, John, Mateusz, Isaac and Krystyn-First Martyrs of Poland, St. Maximilian Maria Kolbe, St. Jack), regional-style buildings were erected at that time (including St. John Baptist, St. John Kanty) as well as buildings representing postmodernism (the church of St. Casimir the Prince, the yet unfinished church of the Divine Mercy) and neo-modernism (the Church of the Blood of Christ). [4]

It seems that the period of dynamic development of sacral architecture in Czestochowa has already passed. By making up for many years of deficiencies in this field, the needs of almost all existing parishes in the city have been met. Only the further territorial expansion of Czestochowa could lead to renewed demand in this field.

Fig. 6 Church of St. Francis of Assisi

5 Urbanistic Aspects of Sacral Architecture in Czestochowa

The oldest sacral buildings in Czestochowa are located in the city center, in the vicinity of the main street, which has been the city axis since the merger of the two centers of Czestochowa and Czestochowka in 1826. The connection of the two former cities, running latitudinally along Virgin Mary Avenue, created the urban axis a modern city called the New City began to shape. In the years 1840–48, the Warsaw-Vienna railway was laid through the city, which enabled further development of the city. In the second half of the nineteenth century, many industrial plants were located mainly on the periphery, gradually being incorporated into the city. In the years 1928–30, due to the development of industry and the expansion of residential districts, the area of the city was further expanded. In these areas, sacral architecture developed, satisfying the spiritual needs of the inhabitants of the new districts. At that time, churches were built in the centers of residential districts, on plots of appropriate size and convenient location [5].

The construction of a large metallurgical plant started in 1951–52 and the new general spatial development plan connected with it resulted in the dynamic development of the city and the construction of numerous large-scale housing estates. All this meant that the borders of Czestochowa were enlarged several times. All the churches built after 1945 were constructed to serve the inhabitants of the growing city. Government authorities effectively blocked the process of building churches, which is why the rapid development of sacral architecture took place only in the 80s. The few churches that were built between 1945 and 1980 were often malevolently located, because the authorities issued permits for construction on the fringes of the residential areas, so that the churches would not be the central points of these new districts.

An example of a church with a very unfavorable location is the church of St. Albert Chmielowski. A construction permit, then for the parish of St. Andrzej Bobola, was issued by the provincial authorities in 1978, after very long efforts to properly locate the place of worship. The voivodship authorities designated a site for construction on Narcyzowa street in Czestochowa. It was a plot of land without the necessary installations, situated in an agricultural area, away from buildings. The pretext for this malicious decision of the then authorities was the planned construction of the North II housing estate. In total, in the near future, the parish was supposed to have 40,000 residents and the church was built for such a number of faithful. However, plans to build the new housing estate were not implemented, and the parish currently has 1700 inhabitants. The church is far too large for this number of parishioners and therefore difficult to maintain. It was built over a period of 13 years and to this day some rooms are still not finished. The conceptual design projected building on a circular plan with a diameter of 40 m, in the technical design the diameter was reduced to 36 m. The designed sacristy behind the presbytery zone and the choir loft, which was supposed to be above it, were not implemented due to financial difficulties.

A further example of a difficult localization is the site of St. Stanislaus Kostka Roman Catholic Church. The church was built on a small plot of land on Dabkowskiego Street in a densely built-up area of the city. One of the walls of the church adjoins a block of flats and the front elevation is situated directly on the frontage of the street. The church building, therefore, does not have parking facilities and, also, there is no possibility for a procession to walk around it. The idea of building a church on such a small plot of land resulted from a growing need to provide the inhabitants of the centre of Czestochowa, where there was a scarcity of undeveloped plots, with a parish. Despite the lack of space next to the church, the area was endowed with some small architectural elements: a cross and a statue of a saint. Additionally, there are two blue spruces planted on either side of the entrance and a lawn with shrubs along the wall, and next to the presbytery there is another silver spruce. Owing to this the whole building seems to be surrounded with greenery and differs from the neighbourhood.

The above objects are examples of how religious architecture was treated in communist times. Many churches, not only in Czestochowa, but all over Poland received plots in unfavorable locations or with an unfavorable configuration and size. This defined the years 1945–1989.

It was not until 1989 that churches were built in all districts where there were no parishes. At present, due to the desire for churches to be close to where their congregations live, very large churches have been allocated to new smaller parishes. The churches built since then are much smaller and easier to maintain. Very large churches, often with an area of over 1000 m^2, are today definitely too large for the reduced number of faithful. Some churches were built too close to existing ones. An example eing the church of St. Kazimierz the Prince, which was built on the same street as the church of St. Adalbert Bishop and the Martyr, which is a very large object with a usable area of over 1000 m^2. Emotional considerations and the strong desire of local residents to have separate parishes decided the mutually impractical location of the two churches.

Roman Catholic church buildings put into use after 1945 are shown on the city map (Fig. 7). The church numbers are as given in Table 1.

Geographic coordinates were found for each of the churches using the GoogleMaps API. The distance matrix was calculated, ordering churches by erection date. A chart was prepared for subsequent erected churches, where the Y coordinate is the distance in a straight line from the nearest erected church. The Least Squares Method trend line has been added.

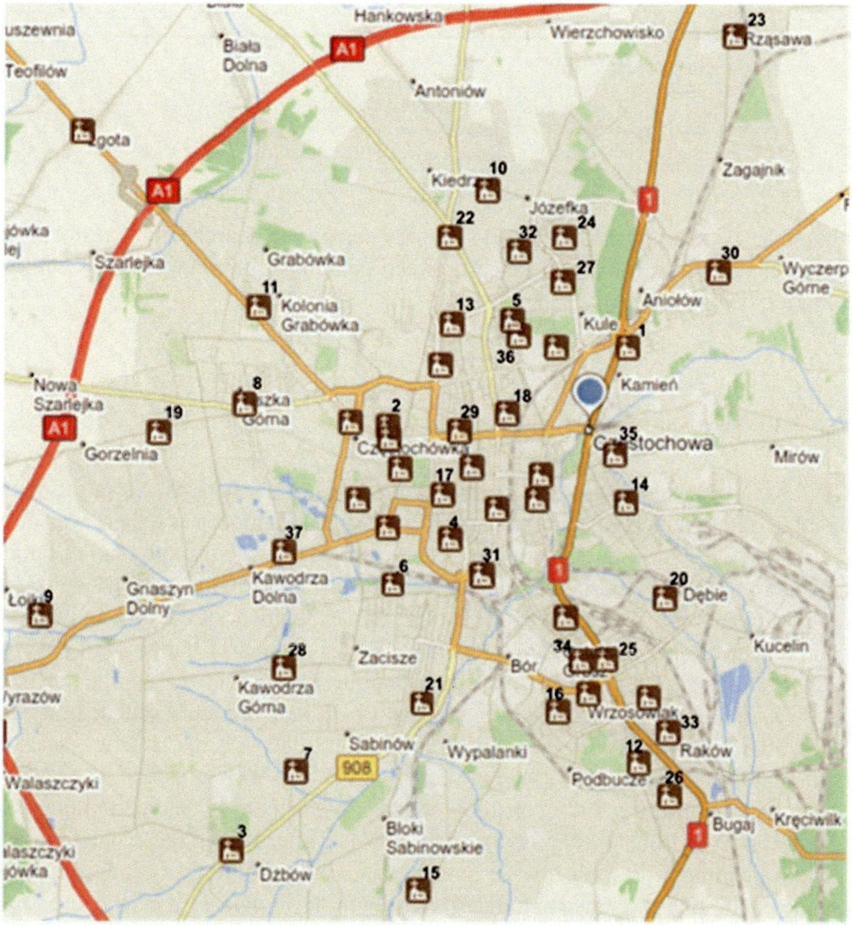

Fig. 7 Contemporary Roman Catholic churches of Czestochowa

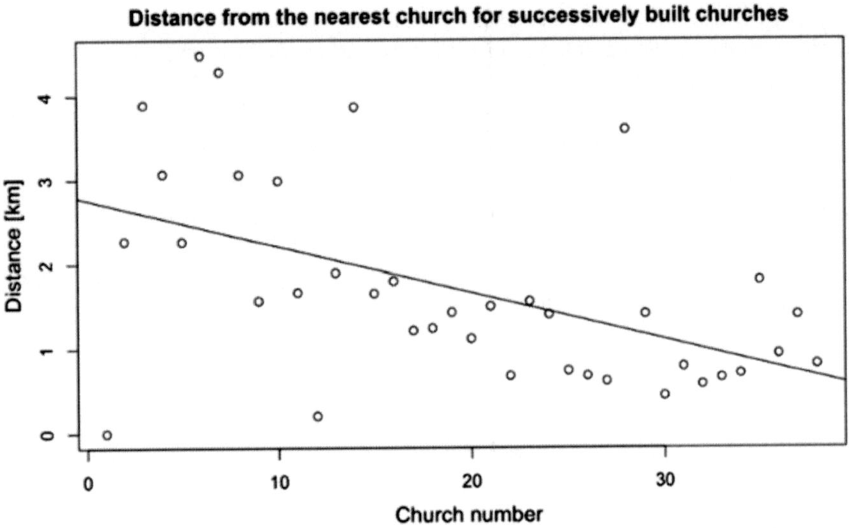

Fig. 8 The relationship between the distance between church buildings over time

As can be seen in the graph (Fig. 8), the density of churches in Czestochowa increases significantly. The distance between churches from an average of almost 3 km fell below 1 km.

6 Summary

Churches in Czestochowa were built in various historical periods. From the fourteenth to the seventeenth century, apart from Jasna Góra monastery, there was only one parish church in the city. In the seventeenth century, two more church buildings were erected. A further expansion of religious construction took place in the nineteenth century, as a result of the industrial development of the city and had its continuation in the early twentieth century. The expansion of the city after World War II was not accompanied by a corresponding increase in the number of religious buildings owing to the anti-church policy of the communist authorities. The largest number of churches were constructed after 1980, and the situation became completely normalized after 1989.

Churches built after 1945 constitute the vast majority of all sacred buildings in Czestochowa. In turn, most of them were established after 1980. These contemporary religious buildings are very diverse both stylistically and in terms of their size. It is characteristic that the largest churches were built at the turn of the 70 s and 80 s, when the first permits for new buildings were issued. This happened because of a fear of a further freeze on permits for new churches. Today, these facilities are too large in relation to current parish needs.

It seems that all needs are currently being met and only further territorial development of the city could result in the construction of new churches. In the years 1945–2005, 37 new churches were built and put into service in Czestochowa, and currently 6 new religious buildings are being constructed: God's Mercy Parish; church of St.Andrzej Bobola; Pentecost parish; church of St. Faustyna Kowalska; Parish of St. Gaspar del Bufalo and St. Melchior Grodziecki [6]. Parish churches were located in all Czestochowa districts, very often (especially those built after 1989) in the district centers on imposing plots, constituting an important urban element. Buildings with a disadvantageous location are a throwback to the difficult communist times. Figure 7 shows a fairly even distribution of churches throughout the city of Czestochowa. Most of them are located near the main communication routes of the city, many large districts (e.g. Tysiąclecie, Północ, Wrzosowiak) have several churches, which means that all the faithful have a place of worship within walking distance, without the need to use a vehicle or public transport.

References

1. Archidiecezja Częstochowska Katalog 2000. Częstochowskie Wydawnictwo Archidiecezjalne REGINA POLONIAE. Częstochowa 2000
2. "Postanowiono załatwić odmownie..." Budownictwo sakralne w diecezji częstochowskiej (1957–1989). Wybór dokumentów. Oddział IPN w Katowicach, Muzeum Częstochowskie, Uniwersytet Rzeszowski. Katowice-Częstochowa-Rzeszów 2008
3. Mazur A (2003) Moje kościoły 1952–2002. Wydawnictwo "Czuwajmy". Kraków
4. Majdowski A Kompozycja przestrzenna powojennych kościołów w Polsce. Magazyn "Ecclesia" nr 5/07
5. Sołkiewicz N–Kos, Czestochowa—the development Phases and their Impact onto Modern Functional and Spatial Solutions [w:]Kreativnij urbanizm: do stolittja misobudivnoi osviti u Lvivs'lij Politechnici (red. Cherkes B., Petryshyn H. Vidavnictvo L'vivsk'koi Politechniki 2014, pp 478–482
6. Repelewicz A The newest sacral buildings in Czestochowa [w:] Architecture in Perspective 11 (red.) Rerinkova M., Juttnerova S., Videcka L. VSB—Technical Univesity of Ostrava. pp 311–313

Representing Built Environments with Digital Tools: Problems with Human Perception

M. Wąsowicz

Abstract The idea of sustainable development in architecture focuses on health care and environmental protection, but not only: Building a healthy human-environment relationship also plays an important role. Demands of sustainable development are fulfilled through ensuring a general sense of well-being, which comprises permanence, safety, and comfort of adapting the built environment to human perceptual abilities and needs. In the twentieth century, research on how the built environment is perceived had gained much attention. Recent studies have applied digital tools in the research process, and sooner than later virtual reality simulations will become an essential tool and an important reference in the science of architecture. Data extracted from virtual models combined with interactive feed-back will be used to learn viewers' understanding of architecture and its surroundings. Until now, researchers have primarily focused on tracing observers' visual dimension of perception and estimating measurable spatial qualities rather than on simulating perceptual experience in real-life contexts. Combined with another problem of human cognitive abilities—that their many different faces reach far beyond visual observations—such an approach to simulation wanders too far from the reality of the build environment. This paper reviews what we known about observing architecture in illusions created in digital environments.

Keywords Digital tool · Virtual model · Perception

1 Introduction

Maintenance and continuity of resources, and the quality of human life—these are the two most common contexts considered along with sustainability, but not the only ones. Another treats ecology as the science of relationships between people

M. Wąsowicz (✉)
Faculty of Civil Engineering and Architecture, Department of Architecture and Urban Planning, Opole University of Technology, Opole, Poland
e-mail: m.wasowicz@po.edu.pl

Z. Zembaty et al. (eds.), *Environmental Challenges in Civil Engineering*, Lecture Notes in Civil Engineering 122,
https://doi.org/10.1007/978-3-030-63879-5_13

and their environment. This reasoning integrally joins biotic organisms and non-biotic elements into a given setting, and people perceive the environment and its various elements from within this environment. Their outlook of the world derives from territorial knowledge along with particular and prompt experience. This leads them to develop a sense of belonging and identity in relation to their surroundings, which sense constitutes a component of widely understood sustainability. People empirically develop connections with their environments through the senses over their intellectual understanding and longitudinal processes of emotional adaptations to the surroundings, established through natural selection during human evolution [1].

To understand relationships between people and their environments, we should determine the association between the physical features of a given setting (e.g., urban landscape or interior space) and human psychological responses. To investigate this topic, we can use digital tools. Such devices are used to evaluate the resemblance between physical architecture and its depiction in virtual reality (VR), with the data obtained that way being used to measure the analogy between the real and the virtual experience. This paper aims to discuss the boundaries of digital representation of architecture, and particularly the criteria of effective visualization in the broad context of human perception. The first part of this paper collates similarities and differences between how the real-life object and the corresponding virtual architecture are perceived. The second part discusses the limitations of sensory simulation in digital technologies, which will help us recognize limitations in generating the sensation of presence in an architectural space.

Virtual reality can be defined as "the representation and simultaneous perception of reality and its physical attributes in an interactive computer-generated virtual environment." [2] Recent research shows a wide range of studies in which VR was applied as a research technique, [3–5] with many applications in the fields of architecture, civil engineering, and environmental planning in terms of visualizing, analyzing, collaboratively designing and redeveloping the built environment. VR has been commonly discussed in the context of building evaluation, examining the experiences of future or current building users, experimenting with spatial cognition, introducing into education, applying in domain-specific expertise, and approaching building design and development through multidisciplinary collaboration based on Building Information Modeling (BIM).

In building evaluation, simulations performed on a digital model can help identify inconsistencies and errors before the construction stage, thereby eliminating the necessity of readjustments and rebuilds during the construction or after it. VR thus helps simulate how future users of a building can view and understand it before its actual construction. It can also simulate how modernizations affect the users' behavior: The related changes in the building are simulated, and the adjusted (still virtual) building is evaluated once more, and additional modernizations can be applied—this process can be repeated any number of times. Data provided from designing and test VR environments can help improve the usability of buildings [5] and enhance their accessibility to people with disabilities or elderly [6]. VR, however, can be implemented also outside of professional architectural practice. It demonstrates its value in architectural education, simulating users' perspectives of students' building designs, and enabling students to experiment with their designs thanks to the ability to see the resulting buildings, even if only in the virtual space [7].

For space reasons, we will not discuss all these applications, however interesting and important they are. Instead, we will focus on issues related to spatial cognition that are crucial to compare virtual and real design environments. This aspect of VR is crucial, because if spatial cognition fails, what we see in a virtual model will unlikely reflect what we would see in reality—a designer's failure.

2 Virtual and Real-World Environments

To accommodate the process of developing virtual buildings to users' perceptual needs, we have to consider human cognition. To learn how virtual reality and the cognitive processes of an individual interact, investigations comparing the two environments—virtual and actual—are crucial. Recent years have witnessed a growing interest in studies correlating user perception of an existing building and of its virtual counterpart [5, 8] Such research usually uses the detailed data of human-computer interaction, indicating where participants maneuver, stop and observe, the records of their eye movements, or even employ brain-computer interfaces (BCIs) and track physiological responses. Often, questionnaire studies support these data with information about the participants themselves as well as their reactions and opinions.

Such environmental comparability is studied in terms of the perception of action and recognition. Three cognitive tests provide necessary information, namely those of (i) the usability of the environments, (ii) strategies for navigation and wayfinding in them, and (iii) spatial memory after changes in spatial arrangement. Slater et al. have recently reviewed the contemporary research on these cognitive processes in the context of virtual reality [4].

Among the three issues, navigation and wayfinding methods stand out, since the related research has not provided conclusive conclusions yet. Some studies have shown that people may be able to approach wayfinding in a similar way in both virtual and real environments [9, 10] Other studies, however, failed to reach such a conclusion [11].

Actually, many environmental comparability studies have led to a conclusion that in virtual reality, people often have distorted perceptions of spatial properties, the estimation of distance and speed being most affected. Johnson [8], Campos [12] and Kuliga [5] reviewed studies that discussed underestimation, inaccuracy and correspondence in distance assessment. Aspects of visual inputs that were identified as important for the accurate perception of absolute distance include the properties of visual graphics (texture and shadows); restrictions to the information from the field of view, including view range; mobility; the accurate perception of eye-height; angular declination from the horizon; and the absence of near-space scaling. Wu [13] emphasized the role of near ground information in scaling far ground. Furthermore, real and virtual environments have been shown to differ in terms of "perceptual-motor coupling," which in simple words means that people differently perceive speed in them [14]. In 1979, Gibson claimed that "as eye-height increases, the magnitude of optic flow decreases," [15] an assumption that helped Campos et al. prove, over thirty years later, that people's eye height affects their speed perception in VR [12].

Aside from cognitive processes, in assessing people's feedback on virtual environments, we need to consider two important factors: affective response and esthetic preferences. Various studies have underlined that participants reported fondness in real built environments [16], assessed architectural spaces generated in the virtual reality lower than their real counterparts, and showed no psychological arousal. We could therefore deduce that affective response is associated with the resemblance of real-world and virtual environment [17]. What can also affect people's feedback is an order in which they experience the two environments. It affects how users environmental valuation and estimate distance [5].

Participants' emotions and cognition are assessed using tools that analyze adjectives used by the participants, like bipolar semantic differential scales. These scientific tools convert words that participants used to describe built spaces studied into indicators of how participants experienced them. Spatial attributes—such as complexity, spaciousness, accessibility and openness—approximately reflect the subjective evaluations of the participants [17].

Such research has various limitations. Users often misinterpret virtual space, and so they experience it in a different way from how they would experience the corresponding real space. Measuring and experiencing a virtual environment strongly depends on the complexity of the devices a VR system consists of, and on the multidimensionality of human perception. Quite often, presentation devices providing visual rendering are combined with supplementary hardware, such as haptic devices, auditory equipment and motion tracking sensors, which aims to deepen participants' multi-sensory experience. Sometimes, however, these devices simply fail, with their representation of real-world characteristics being far from perfection, showing either a distorted reflection of reality or even something entirely unlike [12].

3 Virtual Environment and Human Senses

Human cognition relies upon the stimuli gathered by sensory systems: vision, sound, touch, force, taste, and smell. People accommodate numerous sensory data in order to support a coherent experience of their environments. In addition to cognition, attention also connects stimuli from various sensory processes: vision, hearing, motor function, as well as effort and touch [8]. Because of how people perceive the surrounding environment, it is crucial to adequately and accurately represent authentic sensory impression, in terms of apprehension and the sense of presence, in the virtual environment. In his 1965 seminal article, Ivan Sutherland proposed "the ultimate display" notion, which in essence can be related to the direction in which the technological development of sensory substitution is heading. Sutherland established three characteristics of VR: immersion, interaction, and imagination [18]. Since they are important concepts, let us discuss them one by one.

VR systems can be classified in terms of immersiveness. An immersive VR system enables a person to perceive a visualized object through his natural sensorimotor capabilities, and represents as many aspects of human perception as technically possible. While immersion refers to the technical capabilities of a system, its emotional side is represented by presence. Presence is defined after Minsky [19] as the subjective, psychological state of being in one place while physically being in another. In this process, the recognition of the real environment is transformed to the recognition of the digital environment regardless of the observers' certainty that what they see is not authentic. The illusory sense of place and reality within the virtual environment is what discerns it from alternative media [4].

The second characteristic of VR according to Sutherland is interaction, here representing interaction between a person and the virtual reality. We have to consider a user's experience in using digital tools. Navigation within virtual representations is substantially different from navigation in real built environments. Being virtual not only in name, virtual realities are practically unexplorable without dedicated tools—namely, human-computer interaction (HCI) devices and methods. According to ISO standard "Ergonomics of human-system interaction—Human-centred design for interactive systems," HCI "includes individual preferences, psychological responses, behaviors (etc.) that happen before, during and after an interaction with a product." [20]. In advanced VR systems, interface is substantially different from a stationary computer with external control devices, such as a mouse. In HCI VR systems, it is possible to use one's body to interact with the digital world, instead of employing devices to fulfill autonomous tasks [4]. At times, complex VR systems lead to less behavioral realism, one of reasons being difficulty with controls [5]. This proves that user experience plays a role in VR, perhaps even crucial, at the very least in complex virtual systems. This aspect needs to be investigated, in the context of employing control devices to navigate through virtual reality, aside from the environmental comparability issues. So important an aspect of people's proficiency in virtual navigation, user experience must not be ignored when designing systems for specific applications and when analyzing behavioral data collected from VR studies [12].

The resemblance factor is perhaps the most commonly discussed characteristic of virtual environments, and it is related to the perception of recognition. Sutherland has its own term for it: imagination. The realism of a virtual environment has been suggested to increase with the sense of presence [21]. Beginning with Radford et al.'s seminal an article from 1997, [22] researchers have investigated how the balance between the three key factors defining a successful visualization—namely, abstraction, accuracy and realism—influences the effectiveness of their use. Recently, however, this concept's relevance has been decreasing: These days, visual displays can have the optical resolution of the human eye, and the computing power of modern VR digital devices allows one to map the virtual environment with great accuracy. With current technical possibilities in HCI, what was a must some time ago—a compromise between processing time and the level of detail and complexity—no longer is a must; what was a considerable issue—significant human contribution to the process of creating visualization—no longer is an issue. The changing HCI world called for a change in the conceptualization of VR. In 2003, Ferverda [23] heard this call, which led him to propose applying Margaret Hagens' varieties of realism in computer graphics. These are physical realism, photorealism, and functional realism [24], a classification employing the differentiation of two neural mechanisms: perception for recognition and perception for action.

The following subsections review state-of-the-art of technical development in VR devices. While virtual reality is always linked with visuality, rapid technical development changed that years ago. Immersive virtual environments provide the user with three types of information: visual, auditory, and haptic. Many modern VR systems employ visual devices supported not only by auditory (which was nothing new even twenty or thirty years ago), but also tactile equipment. This is novel—but also powerful. Human-computer interaction through visual and auditory senses offers a great deal of possibilities, but incorporating the sense of touch makes a whole world of difference, opening a new worlds of virtuality.

Less often do VR systems deliver stimuli to sensory systems of smell, and almost never of taste, making them less relevant in experiencing built environments. For this very reason, we will not discuss devices stimulating senses of smell and taste.

Given VR's immersiveness, it is best to include as many sensory systems as possible [4]. Hence, to evaluate VR devices in terms of sensory substitution, we will use the following factors, as critical for this process's effectiveness: wide field-of-view vision, allowing for near-space scaling; stereoscopic display; stereophonic acoustics; motion tracking; and low latency.

4 Vision

In essence, VR systems consist of electronic visual display devices, such as a desktop monitor, a multi-monitor display, a head-mounted display (HMD), and a projection screen, which can take various forms [4]. This list does not include a traditional stationary computer monitor. This is because it is not considered immersive, enabling the user to observe visualizations within the existing surroundings, thereby reducing the impression of presence and self-motion.

A multi-monitor display, used in early studies on VR technology, is an array of monitors that together establish a multiple visual display. Their more up-to-date equivalent is multi monitor setups or video walls, consisting of multiple displays, such as LCD panels, Direct View LED arrays, and Laser Phosphor Displays, all set up together in order to form a singular large screen. Such displays have a significantly wider field of view than standard desktop displays, and they can be mounted along curved surface, in order to provide the user with a more peripheral view. A panoramic projection display comprises an adjusted setting of surfaces and an image projection system. One of the first projection displays was the Cave Automatic Virtual Environments (CAVE), which took a form of a room-sized cube. On its walls and the ground, a projection of the visual scene is displayed, with two slightly different images, which aims to account for the inter-pupillary distance. The user wears stereoscopic LCD shutter glasses providing a 3D display of the environment. In recently developed panoramic displays, projection screens wrap around a viewer in the shape of a dome or curved surfaces. That way the display incorporates the full visual field, allowing the user to actively move within the setup while maintaining visual association to her own body. Her movements are only restricted by the range of the projection screens. Comparing to multi-monitor displays, panoramic projection displays provide significantly more near-space information, which, as already mentioned, is crucial in estimating distance. Likely the only noticeable disadvantage of such displays may arise in terms of the visual representation of the virtual environment: Depending on the display's form, the user's view can be distorted on surface joints [12].

Head-mounted displays (HMDs) track and render head movements and support accurate navigation. This type of display was initially developed for entertainment purposes, but this changed quite soon, with devices such as the Oculus Rift, HTC Vive, Google Cardboard, Samsung Gear VR LeapMotion, Virtuix Omni being currently reasonably priced, but very efficient research tools. Devices worn on a user's head contain either one (monocular HMD) or two small screens that display an image to each eye (binocular HMD). The two images can either be the same or support a stereoscopic display. Currently, HMDs can be applied for both virtual and augmented reality (AR). In the latter, a computer-generated image can be superimposed on a real-world view. HMDs with a nontransparent display only offer limited visual information to the user, who can see what screen devices show, and nothing else. HMDs can be combined with a motion tracking system, which updates the visual image as the user moves her head. Such a possibility compensates for the limited vertical field of view, allowing the user to see a wider range of the virtual space than she would be able to in a fully stationary position [12]. This is not the only limitation this type of display, another being HMD's weight, which can affect distance recognition in virtual environments [25].

In terms of immersiveness, two of the above-mentioned displays strongly support visual stimuli: panoramic displays and head-mounted displays. Both often used as research tools, thcy do differ. IIMDs offer smaller fields of view, and only allow for several participants to communicate in one environment through virtual representations of their bodies. Panoramic projection displays offer more possibilities, some of them supporting full fields of view and providing settings for several participants to retain visual connections to their physique [8].

State-of-the-art technology can support fully immersive visual substitution with the following technological solutions: a full field of view, a stereoscopic display, images of very high-quality, increased mobility within the virtual environment when wearing an HMD, and integration with sound and haptic equipment.

5 Acoustics

Visual information—even in itself a powerful means of exploring virtual realities—can be enhanced by auditory information. In some scenarios this will be more effective than in others, such as in situations in which the sense of hearing can be particularly effective, like in virtual walks through urban or architectural interiors. People are responsive to the temporal, spectral and spatial acoustic stimuli in the built environment. Requiring large amounts of data, this sensitivity hinders realistic simulation of interior acoustics, and so, to transmit auditory environment in a persuasive way, high-quality measures are needed. Acoustic simulation needs to include sounds affecting surface materials and architectural features, such as room volume and its structural properties [2]. So detailed resemblance of sounds does not pertain to all aspects of acoustic simulation, however. Human perception research has revealed that, exposed to visual and acoustic stimuli at the same time, people react to the former much more than to the latter [26]. It is thus the accuracy of visual simulation that makes a difference, and acoustic simulation's less so: if the visuals are accurately conveyed, even less accurate acoustics should do the job [27]. Thus, dissimilarities between real and virtual sound turn out to be less apparent when accompanying visual stimuli. This is not to say, however, that the industry is happy with inaccurate visual sound only because it is less important than the visual display it accompanies. Soon, all modern VR systems exploiting acoustic stimuli will do so with accuracy, searching for the simulation of real environments in all their aspects.

In order to use auditory support, the user has a sound system fixed to her head: headphones. The system can be independent of the visual display, but modern HMD go a step farther, joining the two systems into one whole: they have an in-built audio output device (Fig. 1). This integration is crucial in terms of immersiveness, because acoustic representation directly relates to motion, is interactive, and is determined by the participant's movements [2].

Recent years have brought about the acknowledgment of how important the synchronization of spatial audio reproduction with visual information is for increasing VR immersion. Hansung et al. used the state-of-the-art of VR systems to reproduce a VR environment with acoustic properties based on a real interior scene captured by two 360° cameras. The participants navigated through the captured scene represented as a virtual model. The researchers compared the sound recorded in the actual location with that digitally generated. The participants explained their impressions, which enabled Hansung et al. conclude that digitally generated sound could be of sufficient quality to sufficiently represent real sound in VR [27].

Vorländer discussed the limitations of architectural acoustics in virtual reality. [2] (Problematic back in 2015, when Vorländer wrote his article, not all of them are currently seen as limitations.) Among those he mentioned, two seem particularly important. To map realistic virtual sounds is much more challenging than to generate visual images. The real-time integration and processing of data from motion-tracking equipment, audio hardware, signal processing, and audio replication is difficult and has constraints, because of latencies resulting from computation time. Therefore,

Fig. 1 Wąsowicz 2019. Virtual reality headset HTC Vive with an in-built audio output device and a controller

audio simulations generated using modern tools have to be significantly simplified; or inversely, it is the room geometry for accurate audio computation quality and optimal processing time that must be significantly simplified [2].

In the implementation of acoustic stimuli in BIM technology integrated with virtual reality, also software compatibility plays role. Acoustic simulations require separate spatial virtual models in which objects deriving from a CAD model have to be replaced by a flat surface with adequate acoustic properties. The resulting data describing the performed acoustic stimulation are stored in one specific data format, which adds another limitation to the direct exchange of such data between different applications. In a long-term perspective, this drawback will likely be resolved by the integration of detailed acoustic data into the BIM-based virtual building model [2].

It is possible to support effective auditory substitution if the following technological variables are provided: stereophonic sound; low latency; high sound quality; and motion tracking, to reflect acoustical point of view at the listener's location.

6 Haptics

Apart from visual and acoustic stimuli, people experience their environments through other modalities, such as motor input [28]. In terms of perception of action, motor input has been recognized as crucial for wayfinding and navigation [14]. Research on goal-directed neural mechanism showed that visuals do not have to accurately resemble real-world environment to be effective, with a fair level of abstraction and only medium visual detail being enough for a virtual environment to represent its real-life counterpart [29]. In addition, visual flow has been proven to affect perceptual processes [8]. Human movements can be simulated manually by using an output device such as gamepad, or alternatively through physical movements within a visual display setting or on a treadmill [5].

Systems interfacing with the sense of touch and force are called haptic systems. Virtual reality systems incorporating haptic systems have greater potential in simulating real-world than do those without them. Haptic devices serve two functions: sensing human manipulation, and simulating force signals from the virtual environment and exerting these forces on the user's body; the latter function is called actuation [30]. VR systems exploit the following haptic devices: exoskeletons and stationary devices; gloves and wearable devices; point-sources and specific task devices; and locomotive interfaces. Feedback devices are also developed, including feedback input devices, force feedback devices, and tactile displays [25]. Thanks to the development of such systems, contemporary VR systems make it possible to simulate object manipulation, motion, and force control tasks, such as grasping, touching and walking. Full-body motion tracking makes it possible to represent avatars of both the observer and other people in the virtual environment, thus allowing for visual feedback of self-movements and multi-user interactions [12]. As was the case with the simulation of sounds, these devices still call for additional development, in order to improve their quality.

In terms of touch-to-touch sensory substitution, most devices provide a simple feedback of vibration. This is because it is difficult to obtain haptic technology that is convincing in terms of simulating touch. It is a far more sensitive faculty than both sight and sound because it requires a sensation to be updated as many as 1000 times per second or more, in order to relay a convincing tactile experience. Feedback data of an object's surface geometry or texture can be provided by tactile feedback systems or tactile displays. While haptic interfaces simulate the shape, weight, or compliance of an object, tactile interfaces simulate the object's surface texture. The sensations of contact are generated to the skin, which is capable of responding to several types of physical sensation: vibrations, small-scale shape or pressure distribution, and temperature sensations. State-of-the-art systems generate combinations of these sensations [25].

Haptic sensations acquired through virtual interaction simulate the corresponding interaction between the human and the real environment, but given the sensitivity of human skin and body in general, the virtual feeling is extremely difficult to imitate given today's technology. Virtual environment comparability research, specifically concerning architectural space perception, has not yet studied the sensory substitution of the sense of touch. We do not know much about the topic, and the industry has failed to produce systems that would enable the user to perceive stiffness, roughness and temperature of objects. This is likely the weakest point of VR systems, calling for intensive research on devices that would convey realistic haptic sensations and feedback [25].

7 Conclusions

In the words of Juhani Palaasma,"Architectural space is lived space rather than physical space, and lived space always transcends geometry and measurability." [31]. Taking into account the undefinable features of human perception addressed by Palaasma, it is fundamental to investigate communication and interaction with architectural representation. Essentially, "the mind infers beyond the information given." [32] People are not passive receivers of stimuli coming from their environments, but actively interpret the information they receive that way, and use this interpretation to construct their subjective impression of the surrounding world. Hence representations are never impartial [12].

The perception of an environment is a multidimensional process that employs many factors, such as image recognition and understanding, cognitive restraints and attention, and—when the environment is virtual, represented using digital technology—user experience. Therefore, in creating a compelling virtual environment, the following aspects need to be considered:

1. A level of real-world resemblance, such as optical resolution, a level of detail, parameters of luminance, and contrast range.
2. Specific technological variables providing sensory substitution: stereoscopy, full field of view, stereophonic sound, and full-body motion tracking.
3. Compensation for misinterpretations of the virtual environment, e.g., of distance and speed estimations.
4. Affective response and aesthetic preferences; a person's response depends upon specialized affective reactions to a greater extent than on knowledge-based logical procedures.
5. User experience, including issues such as the ergonomics of equipment, providing behavioral realism in terms of handling controls, and the reduction of equipment physical impact on participant.

Given the pace of technological development, immersive VR will in the foreseeable future accurately simulate experiences of physical reality in terms of graphics, sound, and motion tracking. But the state-of-the-art in haptic technology still does not enable us to perceive virtual environments the same way as we perceive real environments, and so further advances in the field of haptics are needed. As argued by Slater [3], solutions can take the form of direct brain interfaces generating stimuli in those aspects of sensory substitution in which external devices have the greatest limitations. We can suppose that with such new tools, illusions created in digital environments will be truly indistinguishable from reality [4].

References

1. Williamson TJ, Williamson T, Radford A, Bennetts H (2003) Understanding sustainable architecture. Taylor & Francis
2. Vorländer M, Schröder D, Pelzer S, Wefers F (2015) Virtual reality for architectural acoustics. J Build Perform Simul 8(1):15–25
3. Slater M (2014) Grand challenges in virtual environments. Front Robot AI 1:3
4. Slater M, Sanchez-Vives MV (2016) Enhancing our lives with immersive virtual reality. Front Robot AI 3:74
5. Kuliga SF, Thrash T, Dalton RC, Hölscher C (2015) Virtual reality as an empirical research tool—exploring user experience in a real building and a corresponding virtual model. Environ Urban Syst 54:363–375
6. Palmon O, Sahar M, Wiess LP, Oxman R (2006) Virtual environments for the evaluation of human performance: towards virtual occupancy evaluation in designed environments (VOE). In: Proceedings of the 11th International Conference on Computer Aided Architectural Design Research in Kumamoto, Japan, pp 521–528
7. Schneider S, Kuliga S, Hölscher C, Dalton RC, Kunert A, Kulik A, Donath D (2013) Educating architecture students to design buildings from the inside out: experiences from a research-based design studio. In: Kim YO, Park HT, Seo KW (eds) Proceedings of the 9th International Space Syntax Symposium, Seoul, South-Korea. Sejong University, Seoul
8. Johnson A (2009) Visualisation techniques, human perception and the built environment. Northumbria Working Paper Ser Interdisc Stud Built Virtual Environ 2(2):93–103
9. Skorupka A (2009) Comparing human way finding behavior in real and virtual environment. In: Proceedings of the 7th International Space Syntax Symposium, Stockholm, pp 1–7
10. Werner S, Schindler LE (2004) The role of spatial reference frames in architecture: misalignment impairs way-finding performance. Environ Behav 36(4):461–482
11. Haq S, Hill G, Pramanik A (2005) Comparison of configurational, wayfinding and cognitive correlates in real and virtual settings. In: 5th International Space Syntax Symposium. Delft University of Technology, Techne Press, The Netherlands
12. Campos J, Nusseck H-G, Wallraven C, Mohle BJ, Bülthoff HH (2007) Visualization and (Mis)Perceptions in Virtual Reality, Tagungsband 10. Workshop Sichtsysteme, pp 10–14
13. Wu B, Ooi T-L, He Z (2007) Perceiving distance accurately by a directional process of integrating ground information. Nature 428:73–77
14. Banton T, Stefanucci J, Durgin F, Fass A, Proffitt D (2005) The perception of walking speed in a virtual environment. Presence Teleoperators Virtual Environ 14(4):394–406
15. Gibson JJ (1986) The ecological approach to visual perception. Psychology Press
16. Westerdahl B, Suneson K, Wernemyr C, Roupe M, Johansson M, Allwood CM (2006) Users' evaluation of a virtual reality architectural model compared with the experience of the completed building. Autom Constr 15:150–165
17. De Kort Y, Ijsselsteijn W, Kooijman J, Schuurmans Y (2003) Virtual laboratories comparability of real and virtual environments for environmental psychology. Presence 12:360–373
18. Sutherland IE (1965) The ultimate display, multimedia: from wagner to virtual reality, pp 506–508
19. Minsky M (1980) Telepresence. Omni 2(9):45–52
20. ISO DIS 9241 – 210, Tech. rep., International Organization for Standardization (2010)
21. Witmer BG, Singer MJ (1998) Measuring presence in virtual environments: a presence questionnaire. Presence 7(3):225–240

22. Radford A, Woodbury R, Braithwaite G, Kirkby S, Sweeting R, Huang E (1997) Issues of abstraction, accuracy and realism in large scale computer urban models, Proc. CAAD Futures '97. Springer, Berlin, pp 679–690
23. Ferwerda JA (2003) Three varieties of realism in computer graphics, human vision and electronic imaging VIII, vol 5007. International Society for Optics and Photonics SPIE 5007, pp 290–297
24. Hagen M (1986) Varieties of realism. Cambrige University Press
25. Bar-Cohen Y, Breazeal CL (2003) Biologically inspired intelligent robots. SPIE Press
26. Bailey W, Fazenda B (2018) The effect of visual cues and binaural rendering method on plausibility in virtual environments, Audio Engineering Society Convention 144. Audio Engineering Society
27. Hansung K, Remaggi L, Jackson P, Hilton A (2019) Immersive spatial audio reproduction for VR/AR using room acoustic modelling from 360 images. In: IEEE 26th. Conference on Virtual Reality and 3D User Interfaces (VR)
28. Rieser JJ, Pick JHL, Ashmead DH, Garing AE (1995) Calibration of human locomotion and models of perceptual-motor organization. J Exp Psychol Hum Percept Perform 21:480–497
29. Ruddle R (2006) For efficient navigational search, humans require full physical movement but not a rich visual scene. Psychol Sci 17:460–465
30. Dangxiao W et al (2019) Haptic display for virtual reality: progress and challenges. Virtual Reality Intell Hardware 1(2):136–162
31. Pallasmaa J (2005) The eyes of the skin: architecture and the senses. Wiley-Academy
32. Bruner J (1957) Contemporary approaches to cognition. Harvard University Press

BIM in Polish Public Tenders—Analysis of Selected Tender Procedures

T. Nowobilski, M. Sawicki, and M. Szóstak

Abstract BIM technology is increasingly used in the Polish construction industry, as well as by public procurement entities. The current Act on Public Procurement Law permits the use of tools of electronic modelling of building data in public investments, which can be understood as BIM technology. The purpose of this article is the quantitative analysis of selected Polish tender procedures in which the contracting authority has indicated the obligation to use BIM technology in the subject of the contract. The analysis included, among others, the subject of the contract, conditions for participating in the procedure, and the criteria that were used to select the best offer. The subject of the article involved the analysis of 20 selected Polish tender procedures. The analysis showed that BIM technology is increasingly required by procurement entities during public procurements. In addition, there can be seen to be a continuous increase in the awareness among contracting authorities concerning BIM technology. In tendering procedures, it is more and more required from the bidders to have knowledge about BIM technology, as well as experience in using it. These aspects are willingly used by procurement entities as a non-price criterion for the evaluation of bids. According to the authors, a correctly prepared investment project, which in the case of public procurements should include the development of the correct tender documentation that includes, among others, the specification of essential terms of the contract and information requirements of the awarding entity with regards to BIM (so-called EIR), is important for the appropriate implementation of the investment when using BIM technology.

Keywords BIM technology · Public procurement · Tender procedure · Construction industry · EIR

T. Nowobilski · M. Sawicki · M. Szóstak (✉)
Faculty of Civil Engineering, Department of Building Engineering, Wroclaw University of Science and Technology, Wrocław, Poland
e-mail: mariusz.szostak@pwr.edu.pl

Z. Zembaty et al. (eds.), *Environmental Challenges in Civil Engineering*, Lecture Notes in Civil Engineering 122,
https://doi.org/10.1007/978-3-030-63879-5_14

1 Introduction

The acronym of the word "BIM" can be understood in different ways [1]. The first development of this word defines "BIM" as a building information model, i.e. a digital description of the physical and functional properties of a building, which acts as a source of knowledge and all data about the object. This model is fully available to participants of the investment process and can provide a reliable basis for making decisions during its life cycle, from the first concept to the demolition of the building [2].

"BIM" can also be understood as a process of the creative generation and use of data concerning a building, as well as its design, construction and exploitation during full operation—i.e. the modelling of information about a building. In this process, all interested investment participants have access to the same information at the same time through, among others, the interoperability of technology platforms [3].

Finally, "BIM" involves the management of building information, and thus the organization and control of investment processes, by using the parameters of the digital building model in order to exchange information on assets throughout the entire investment cycle. Due to visual communication using three-dimensional objects, it is possible, among others, to early recognize the possibilities of sustainable, effective, interdisciplinary and interactive design; to conduct inspections during construction on a building site; or to update documentation to the actual state by including "design changes during construction and operation" [4].

To sum up, the acronym "BIM" should be comprehensively understood as a process of creating and managing a building's data throughout its entire life cycle. At the core of BIM technology is software that is used for three-dimensional (3D) dynamic building modelling in real time. Therefore, each model, which is developed in accordance with BIM technology, includes geometric information such as the geometry of individual building elements and the spatial relations between them, as well as non-geometric information such as material properties, prices of individual building components, etc. It should be emphasized that BIM does not only apply to the design stage of a building object, but also to its full life cycle (planning, designing, implementation, use/management, demolition) [5].

2 BIM in Public Procurement

Although BIM technology is more and more available to investors, designers and contractors of construction works, the construction industry is still showing low maturity when using this technology [6]. Polish experience to date shows that there are only a few comprehensive and full applications of BIM technology. On the other hand, there are many more investments using selected tools and elements of this technology.

The experience of other countries when implementing BIM technology indicates that one of the ways to use it in public tenders may be through the statutory obligation for designers and work contractors to use this technology [7]. It should be remembered, however, that the obligatory adoption of BIM technology when implementing an investment in the public sector requires a change in existing practices and existing guidelines regarding the preparation and conduct of tender procedures [8, 9]. The adaptation of the currently applicable provisions in the field of the Public Procurement Law, the development of BIM standards, the conducting of pilot projects, as well as the creating of support for awarding entities and contractors are also important [10].

A milestone in the introduction of BIM technology in Polish public procurement was the amendment to the Public Procurement Law from 2016 [11]. In the amendment to the Act, in Article 10, point e, the following provision was introduced: "in the case of procurement for construction works or design contests, the awarding authority may require the use of electronic modelling tools for building data, or other similar tools". This provision is in line with the requirements of Directive 2014/24/EU of the European Parliament and Council from February 26, 2014 regarding public procurement, which repealed Directive 2001/18/EC, and which in Article 22, point 4 permits the use of electronic modelling tools—"for public works contracts and design contests, Member States may require the use of specific electronic tools, such as building information electronic modelling tools or similar" [12]. The latest Public Procurement Law from September, 2019, which according to [13] will enter into force on January 1, 2021, modified the above provision and expanded the scope of relevant tools—Article 69, point 1: "in the case of procurement for construction works or design contests, awarding entities may require the preparation and submitting of offers or contest entries using electronic building modelling tools or other similar tools that are not generally available." [14].

It should be emphasized that national regulations, which have been in force for several years, allow the use of BIM technologies by contracting authorities for the implementation of investments related to the execution of construction works. The best example confirming this possibility can be seen to be already completed or ongoing public investments that use BIM technology, some of which are presented later in the article. However, special attention should be paid to the technical aspect of the implementation of construction investments based on this technology. Not without significance is the awareness of contracting entities and their adequate preparation for the implementation of such an investment project, and also the development of correct tender documentation—including the specification of the essential terms of the contract (the so-called SETC).

When developing the Specification of the Essential Terms of the Contract (SETC) for a given project, the implementation of which is to be based on BIM technology (at the design and/or execution stage), it is important for the contracting authority to precisely specify the requirements for the future contractor in terms of BIM technology. For this purpose, Employer's Information Requirements (EIR) should be developed [15]. Typical information requirements of the contracting authority consist of:

- technical requirements, including: expectations of the contracting authority regarding: the used software, data exchange formats, levels of detail, training, etc.;
- management requirements, including: data exchange standards, data security, coordination and detection of collisions, IT system performance, etc.;
- commercial requirements, including: contractor's competences, the main goals of the contracting authority, which should be achieved with the use of BIM technology when implementing the investment, etc.

The purpose of the article is a quantitative analysis of selected Polish tender procedures in which the contracting authority in the subject of the contract indicated the obligation to use BIM technology. The analysis included, among others, the subject of the contract, conditions for the participation in the procedure, and the criteria that were used to select the best offer.

3 Selected Public Procurement Implemented Using BIM Technology

In order to determine the number of public procurements in Poland that were based on BIM technology, an online version of the public procurement search engine was used: "Supplement to the Official Journal of the European Union"—a portal dedicated to European public procurement [16]. As a result of searching for the term "BIM" with regards to current and archival procurements in Poland in the years 2014–2019, 47 records (procurement notices) were obtained. From this number, 20 tender procedures were selected for further analysis.

For each analysed public procurement, detailed information was provided regarding:

- the subject of the order,
- conditions for participating in the proceedings regarding the required knowledge and experience of companies in the field of BIM (potential contractors), and also the knowledge and experience of the required staff in the field of BIM,
- SETC requirements regarding the execution of the contract on the basis of BIM technology,
- the criteria used to select the best offer.

The following public procurements were analysed in detail:

1. the construction of the Jozef Pilsudski Museum complex in Sulejówek (2014)—the design and execution of construction works based on the Functional-Utility Program [17],
2. a building of the District Court in Katowice (2015)—carrying out an architectural inventory of the historic building in the BIM model [18],

3. a building of the Academy of Fine Arts of Jan Matejko in Cracow (2015)—preparation of project documentation for the thermomodernization of the buildings [19],
4. the extension of the multi-specialist voivodeship hospital in Gorzow Wielkopolski (2016)—preparation of multi-branch, comprehensive project documentation (building and executive documentation) [20],
5. revitalization of a tenement house on Jana Pawel II Street in Mikolow (2016)—preparation of project documentation [21],
6. the construction of a skiing centre in Szklarska Poreba (2016)—execution of construction and executive projects, as well as managerial and technical activities [22],
7. the reconstruction and extension of the facilities of the Institute of Heat Engineering in Warsaw (2016)—preparation of design documentation for the reconstruction and extension of the facilities [23],
8. the construction of a dormitory in Deblin (2016)—preparation of multi-branch project documentation based on BIM technology, and also the preparation of cost estimate documentation [24],
9. the construction of the Ministry of Internal Affairs and Administration building on Stefana Batorego Street in Warsaw (2016)—preparation of project documentation for the construction of the building [25],
10. the reconstruction of the Ministry of Internal Affairs and Administration building on Domaniewska Street in Warsaw (2016)—reconstruction of the building in the "design and build" system, i.e. the development of a complete, multi-branch project-cost documentation, supervision over the implementation of the project, and execution of construction works [26],
11. the extension of the Thermal Waste Conversion Installation in Warsaw (2017)—design and extension of the installations [27],
12. the extension and the execution of the superstructure of the AGH University of Science and Technology building in Cracow (2017)—execution of multi-branch construction and executive projects, bills of quantities, investor cost estimates and the Specifications for Work Accomplishment and Acceptance [28],
13. the construction of a paediatric hospital in Poznan (2018)—implementation of the BIM model and verification of both the coordination and bill of quantities of executive projects [29],
14. power infrastructure facilities (2018)—execution of BIM models of the proposed power infrastructure facilities [30],
15. the construction of the headquarters of the Poviat Police in Pila (2018)—preparation of technical documentation, cost estimates and technical specifications for the performance and acceptance of construction works in the BIM 5D Level 3 formula (Building Information Modelling) [31],
16. the construction of a skiing centre in Szklarska Poreba (2018)—performance of construction works [32],
17. the reconstruction and renovation of a hospital in Ustrzyki Dolne (2018)—development of project documentation for the extension, reconstruction and renovation of the building [33],

Table 1 List of analysed tender procedures—part 1/2 (own elaboration)

Tender procedure No:	1	2	3	4	5	6	7	8	9	10
The subject of the order	Occurrence of a given aspect in the tender procedure									
Inventory	–	√	–	–	–	–	–	–	–	–
Designing	√	–	√	√	√	√	√	√	√	√
Execution	√	–	–	–	–	–	–	–	–	√
Knowledge and experience of the company in the field of BIM	√	–	–	–	–	–	–	√	√	√
Knowledge and experience of the staff in the field of BIM	√	–	–	–	–	√	–	–	√	√
Records regarding BIM requirements included in the SETC	–	√	√	√	√	√	√	√	√	√

18. the design and construction of the beltway of Zator (2018)—design and construction of the beltway along national road No. 28 [34],
19. the renovation and reconstruction of the Regional Blood Donation and Blood Treatment Centre in Opole (2019)—preparation of project documentation using BIM for the renovation and reconstruction of the building [35],
20. the design and construction of the Biology Center in Wroclaw (2019)—design and construction of the Experimental Animal Infections Center with its equipment [36].

Tables 1 and 2 present the summary of the 20 above-mentioned tender procedures with information about: the subject of the order, the conditions for participating in the procedure with regards to the required knowledge and experience of the company in the field of BIM, the required knowledge and experience of the staff in the field of BIM, and also SETC requirements regarding the implementation of the order in BIM technology. In turn, Tables 3 and 4 contain the percentage share of individual criteria for selecting the best bid.

4 Detailed Analysis of Tender Procedures

Figure 1 shows the scope and the relationships between the subjects of the contract for the 20 analysed tender procedures, while Fig. 2 shows the relationships between the required knowledge and experience of the company and its staff in the field of BIM, as well as the BIM requirements contained in the SETC.

Based on the detailed analysis of the 20 selected Polish tender procedures, it was concluded that:

- the subject of 18 tender procedures was the development of project documentation for the planned construction works, which accounts for 90% of all the analysed

Table 2 List of analysed tender procedures—part 2/2 (own elaboration)

Tender procedure No:	11	12	13	14	15	16	17	18	19	20	Total number of occurrences
The subject of the order	Occurrence of a given aspect in the tender procedure										
Inventory	–	–	–	–	–	–	–	–	√	–	2
Designing	√	√	√	√	√	–	√	√	√	√	18
Execution	√	–	–	–	–	√	–	√	–	√	6
Knowledge and experience of the company in the field of BIM	–	–	√	–	√	–	√	√	√	–	9
Knowledge and experience of the staff in the field of BIM	√	–	√	√	√	√	–	√	√	–	11
Records regarding BIM requirements included in the SETC	–	√	√	√	√	√	–	√	√	√	17

Table 3 Percentage share of individual selection criteria for the best offer for the analysed tender procedures—part 1/2 (own elaboration)

Tender procedure No:	1	2	3	4	5	6	7	8	9	10
Criteria for choosing the best offer	Percentage share of individual selection criteria for the best offer [%]									
Offer price	46	100	80	80	80	60	40	50	55	80
Efficient use of energy	46	0	0	0	0	0	0	10	0	0
Deadline for completing the subject of the contract	4	0	0	0	0	0	20	40	0	20
Implementation of the project in the BIM system/BIM methodology	0	0	20	0	20	40	40	0	45	0
Innovation of used solutions/quality of solutions	4	0	0	18	0	0	0	0	0	0
Date of payment	0	0	0	2	0	0	0	0	0	0
Guarantee period	0	0	0	0	0	0	0	0	0	0
Staff experience	0	0	0	0	0	0	0	0	0	0

Table 4 Percentage share of individual selection criteria for the best offer for the analysed tender procedures—part 2/2 (own elaboration)

Tender procedure No:	11	12	13	14	15	16	17	18	19	20	Total number of occurrences
Criteria for choosing the best offer	Percentage share of individual selection criteria for the best offer [%]										
Offer price	60	60	60	100	60	60	60	40	60	60	20
Efficient use of energy	20	0	0	0	0	0	0	0	0	0	3
Deadline for completing the subject of the contract	0	0	40	0	0	0	0	0	0	0	5
Implementation of the project in the BIM system/BIM methodology	0	15	0	0	40	30	20	60	0	10	11
Innovation of used solutions/quality of solutions	0	0	0	0	0	0	0	0	10	0	3
Date of payment	0	0	0	0	0	0	0	0	0	0	1
Guarantee period	10	0	0	0	0	0	0	0	0	0	1
Staff experience	0	25	0	0	0	10	20	0	30	30	5

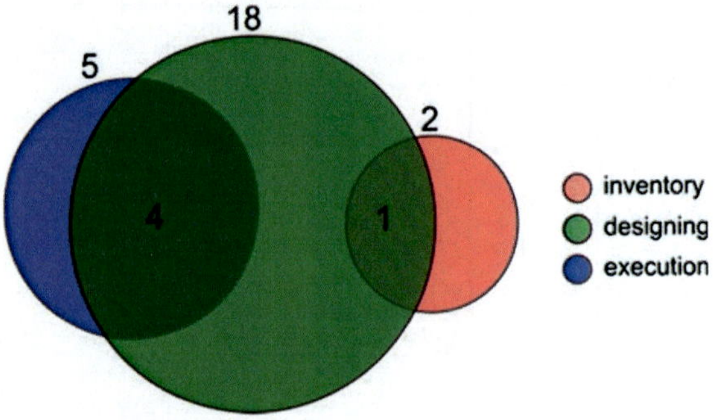

Fig. 1 The subjects of the contract (own elaboration)

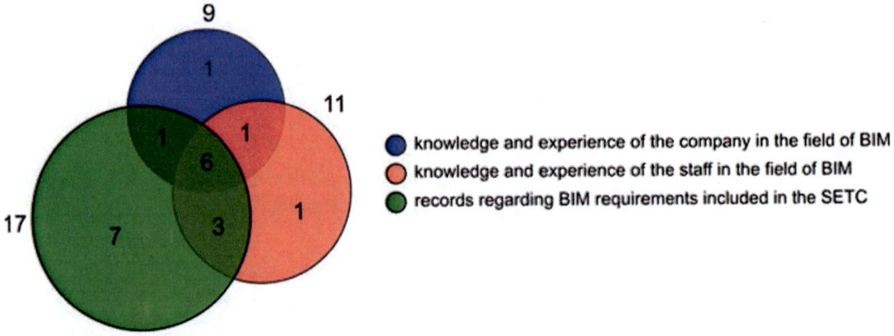

Fig. 2 The required knowledge and experience of the enterprise and staff, as well as the SETC requirements in the field of BIM (own elaboration)

procedures, while the subject of 6 procedures was the performance of construction works, which represents 30% of all the analysed procedures,

- in 5 tender procedures, the contracting authority decided to use the "design and build" formula, i.e. the preparation of project documentation and the execution of construction works on the basis of this documentation, which constitutes 25% of the analysed tender procedures,
- provisions regarding the detailed requirements of the contracting authority concerning the fulfilment of the contract using BIM technology were specified in 17 tender procedures (85% of all analysed procedures); 9 tender procedures required the knowledge and experience of the bidding enterprise (45%); and 11 procedures required knowledge and experience of key personnel (e.g. designers, BIM manager) in the field of BIM (55%),
- in 7 tender procedures, the contracting authority did not require tenderers to have knowledge and experience in BIM technology with regards to an enterprise and the staff carrying out the project, despite the fact that the requirements for the implementation of the subject of the contract indicated taking into account detail requirements regarding BIM technology (e.g. the determination of the levels of

detail, the development of the BIM Implementation Plan, or the application of the appropriate file exchange format).

5 Analysis of the Criteria for Selecting the Best Offer

Figure 3 presents the percentage share of the price criterion in the 20 analysed tender procedures, while Fig. 4 presents a bar chart that illustrates the number of occurrences of individual identified non-price evaluation criteria of offers, which are ranked from the maximum to minimum (the so-called Pareto chart), as well as the cumulative percentage share of subsequent criteria (in the form of the so-called Lorenz curve). In order to select the most important criteria, Pareto-Lorenz analysis and the Pareto principle 80/20 were used. According to this principle, the set of relevant criteria accounts for 80% of all the criteria [37].

Based on the analysis of 20 tender procedures, it was found that:

- in 11 tender procedures, the non-price evaluation criterion was used while taking into account the aspect of the execution of a contract using BIM technology, which constitute 55% of all the analysed procedures,
- in the remaining 9 proceedings, non-price selection criteria referred to, among others, the efficient use of energy (3 procedures), the shortening of the deadline for completing the subject of the contract (5 procedures), the innovation and quality of solutions not related to BIM technology (3 procedures), the payment period (1 procedure), and the warranty period (1 procedure),
- the experience of key personnel in the field of knowledge of BIM technology as a non-price criterion for bid evaluation appeared only in 5 tender procedures,

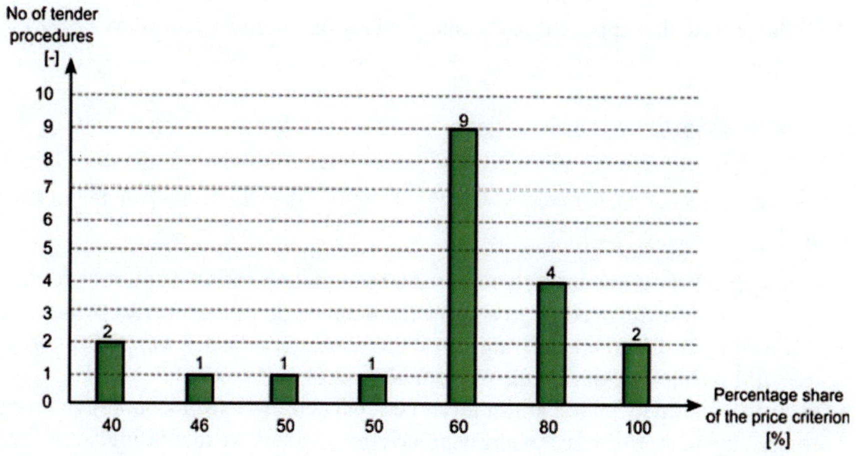

Fig. 3 Percentage share of the price criterion in the 20 analysed tender procedures (own elaboration)

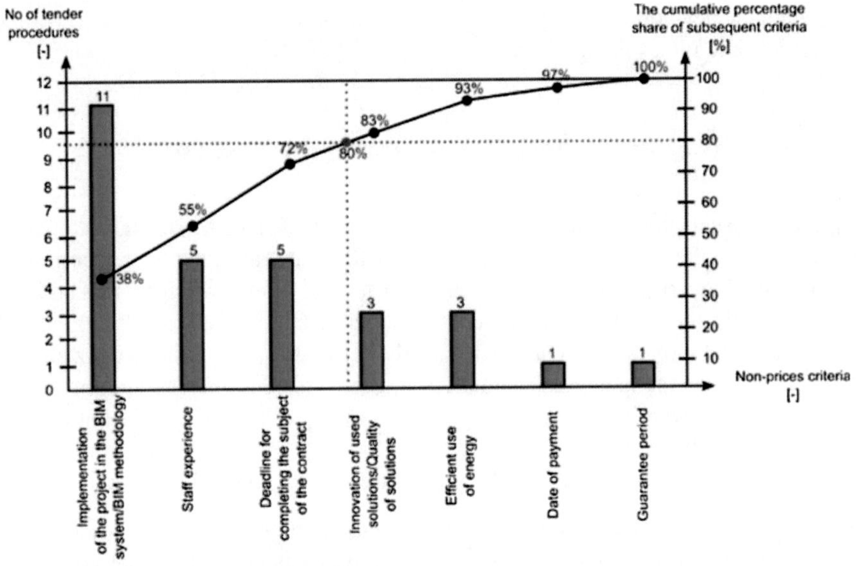

Fig. 4 Pareto-Lorenz chart for non-price evaluation criteria of bids (own elaboration)

- only in 4 proceedings was the criterion of the lowest price no more than 50%, and in 6 cases this criterion was 80% or more,
- as a result of the Pareto-Lorenz analysis, it was found that the most common and significant non-price criteria for the offers included the implementation of a project in the BIM system/BIM methodology (approx. 38%), staff experience in the field of BIM (approx. 18%), and also the deadline for completing the subject of the contract (approx. 18%). The indicated criteria constitute approx. 72% of all the criteria that appeared in the analysed set of 20 tender procedures.

6 Conclusions

As a result of the conducted analysis of 20 selected tender procedures related to BIM technology, it was found that:

- BIM technology is increasingly required by contracting entities in public procurement. This is evidenced by the growing number of procedures in which the contracting authority in the subject of the contract indicates the obligation to use BIM technology during the implementation of the project.
- there is a continuous increase in awareness concerning BIM technology among contracting authorities. In tendering procedures, bidders are increasingly required to possess, among others, knowledge of BIM technology and experience in its use. These aspects are willingly used by contracting entities as a non-price criterion for the evaluation of bids. In the analysed group, non-price bid evaluation criterion

related to BIM technology occurred in 11 out of 20 proceedings, which is 55% of the analysed set of proceedings.

- an appropriate preparation of the investment project is important for the correct implementation of the investment when using BIM technology. In the case of public procurement, the preparation particularly includes the development of correct tender documentation that involves, among others, the specification of essential terms of the contract, and also information requirements of the contracting authority with regards to BIM (the so-called EIR). In the above documentation, the contracting authority should precisely specify the requirements regarding, among others, the required data exchange format, the expected degree of detail of the model in its various phases, the goals that are intended to be achieved using BIM technology during the implementation of the investment, etc.

References

1. Kasznia D, Magiera J, Wierzowiecki P (2018) BIM w praktyce. Wydawnictwo Naukowe PWN, Warszawa
2. Attenni M (2019) Informative models for architectural heritage. Heritage 2:2067–2089
3. Enshass M, Al Hallaq K, Tayeh B (2019) Limitation factors of building information modeling (BIM) implementation. Open Constr Build Technol J 13:189–196
4. Doan D, Ghaffarianhoseini A, Naismith N, Zhang T, Ehman A, Tookey T, Ghaffarianhosein A (2019) What is BIM? A need for a unique BIM definition. MATEC Web Conf 266:05005
5. Kovacs A, Micsik A (2019) Method for evaluating a building information model. Period Polytech Civ Eng 63:541–549
6. Azhar S (2011) Building information modeling (BIM): trends, benefits, risks, and challenges for the AEC industry. Leadersh Manag Eng 11:241–252
7. Porwal A, Hewage K (2013) Building information modeling (BIM) partnering framework for public construction projects. Autom Constr 11:204–214
8. Piotrowski T (2019) BIM - obowiązek czy możliwość stosowania? BIM w prawie zamówień publicznych - cz. 1, Inżynier Budownictwa 5:40–41
9. Piotrowski T (2019) Interoperacyjność. BIM w prawie zamówień publicznych - cz. 2, Inżynier Budownictwa 7(8):56–58
10. Anger A, Lisowski B, Piwkowski W, Wierzowiecki P (2015) Ogólne założenia procesu wdrażania bim w realizacji zamówień publicznych na roboty budowlane w Polsce, Przegląd Budowlany 10:6–9
11. Act of 22 June 2016 amending the act—Public procurement law and some other acts (Law Gazette, 2016, it.1020, as amended)
12. Directive 2014/24/EU of the European Parliament and of the Council of 26 February 2014 on public procurement, repealing Directive 2001/18/EC
13. Act of 11 September 2019 Provisions introducing the act—Public Procurement Law (Law Gazette, 2019, it.2020, as amended)
14. Act of 11 September 2019 Public Procurement Law (Law Gazette, 2019, it.2019, as amended)
15. PAS 1192-2:2013 Specification for information management for the capital/delivery phase of construction projects using building information modelling
16. https://ted.europa.eu. Access date 13 Jan 2020
17. https://www.e-bip.org.pl/muzeumpilsudski/1241. Access date 13 Jan 2020

18. http://archiwumbip.katowice-zachod.sr.gov.pl/bip/zamowienia/ponizej_14/258.html. Access date 13 Jan 2020
19. https://www.asp.krakow.pl/index.php/pl/aktualnoci/zamowienia-publiczne/przetargi/2229-2015-10-09-08-59-25. Access date 13 Jan 2020
20. https://ted.europa.eu/udl?uri=TED:NOTICE:85078-2016:TEXT:PL:HTML&src=0. Access date 13 Jan 2020
21. http://bip.mikolow.eu/?a=3982. Access date 13 Jan 2020
22. https://ted.europa.eu/udl?uri=TED:NOTICE:107106-2016:TEXT:PL:HTML&src=0. Access date 13 Jan 2020
23. https://www.zamowienia.pw.edu.pl/wykaz/szczegoly.php?id=4729. Access date 13 Jan 2020
24. https://ted.europa.eu/udl?uri=TED:NOTICE:250715-2016:TEXT:PL:HTML. Access date 13 Jan 2020
25. https://ted.europa.eu/udl?uri=TED:NOTICE:262381-2016:TEXT:PL:HTML. Access date 13 Jan 2020
26. http://www.przetargi.egospodarka.pl/159755_Przebudowa-budynku-Ministerstwa-Spraw-Wewnetrznych-i-Administracji-przy-ul-Domaniewskiej-36-38-w-Warszawie-w-systemie-projektuj-i-buduj_2016_2.html. Access date 13 Jan 2020
27. https://www.competitionline.com/de/ausschreibungen/266008. Access date 13 Jan 2020
28. http://przetargi.propublico.pl/OgloszeniaSzczegoly.aspx?id=37892. Access date 13 Jan 2020
29. http://przetargi.propublico.pl/OgloszeniaSzczegoly.aspx?MasterPage=EmptyMasterPage&id=41382. Access date 13 Jan 2020
30. https://www.pse.pl/przetargi/przetargi-niepubliczne/przetargi-niepubliczne-aktualne?safeargs=61727469636c6549643d323033383138303832. Access date 13 Jan 2020
31. http://bip.poznan.kwp.policja.gov.pl/KWP/zamowienia-publiczne/archiwum-sprawy-zreali/rok-2018/28046,ZZP-2380-282018-Opracowanie-dokumentacji-projektowej-w-formule-BIM-5D-Level-2-Bu.html. Access date 13 Jan 2020
32. https://platformazakupowa.pl/transakcja/181164. Access date 13 Jan 2020
33. http://www.bip.bieszczadzki.pl/strona-876-opracowanie_dokumentacji_projektowej_na.html. Access date 13 Jan 2020
34. https://www.gddkia.gov.pl/pl/2783/zamowienia-publiczne-rozstrzygniete,nr_ref-,data_postepowania_od-,data_postepowania_do-,nazwa-Zaprojektowanie%20i%20budowa%20obwodnicy%20,opis-,zakres-,oddzial-306,typ_progu-,sortuj_po-data_postepowania,sortuj_jak-DESC. Access date 13 Jan 2020
35. https://www.rckik-opole.com.pl/przetargi-rckik-opole-1-1.html. Access date 13 Jan 2020
36. https://www.upwr.edu.pl/przetargi/49192/r0ap0000_271_17_2019.html. Access date 13 Jan 2020
37. Hoła A, Sawicki M, Szóstak M (2018) Methodology of classifying the causes of occupational accidents involving construction scaffolding using pareto-lorenz analysis. Appl Sci 8:48

Delayed Building Schedule Control Using Milestones

D. Przywara and A. Rak

Abstract A feasibility study is a measure of the quality and reliability of construction schedules. In the course of construction projects, there are a number of external factors that generate time delays and cost variances. Their detailed monitoring and identification allows managers to implement improvement measures. The purpose of the research is to build a decision model that controls the distribution of the resulting time and cost variances identified on the given day of the work registration report. The works assortment processes burdened with the variances in the preceding activities, foreseen to be performed after the selected inspection day were subjected to the improvement measures. Delayed schedule control using milestones allows for the division of time slips by setting priority working deadlines. These deadlines initially eliminate time reserves, making all activities critical.

Keywords Time anchors · Milestones · Delayed schedule

1 Introduction

In the production system, there is a flow of materials gradually changing their form and properties, as well as information and decision-making processes, which are the basis for controlling material flow. There are products at the output of such a system with specific technical, functional and aesthetic features, that meet the specific needs of the target audience of customers [1].

The decisive factor for accurate budget planning in a separate budgeting entity is the way in which orders related to this entity are linked to cost planning during budgeting periods [2]. It allows to precisely identify each type of costs and relate it to the right place and time of its creation.

Modern decision-making process is the problem of choosing the best decisions from the group of available options. All decisions regarding individual elements of

D. Przywara (✉) · A. Rak
Faculty of Civil Engineering and Architecture, Department of Civil Engineering and Construction Processes, Opole University of Technology, Opole, Poland
e-mail: d.przywara@po.edu.pl

Z. Zembaty et al. (eds.), *Environmental Challenges in Civil Engineering*, Lecture Notes in Civil Engineering 122,
https://doi.org/10.1007/978-3-030-63879-5_15

assets must include systematic analysis of all elements of risk, possible ways of determining its degree, and possible ways of controlling it [3].

The selection of production control methods depends on many factors [4]—above all on the degree of production repeatability, as well as the level of detail of available data, the detail of the records kept, the use of planning and recording documentation, the degree of use of computer techniques and the prevailing organizational culture in the enterprise. Controlling the production flow requires the use of appropriate source data and planning norms as well as the preparation of an appropriate production plan. Each enterprise maintains the appropriate data sets containing information on production cycles from previous seasons.

The economic analysis of the project is an iterative process regarding the moment of its launch (pre-investment phase) and periodic assessments during the project implementation, when changes in the surrounding environment and level of knowledge, or finding serious discrepancies in relation to what was foreseen, lead to a review of previous operational strategies [5].

Cost accounting is the basic unit used to monitor and control the costs of projects. It has a hierarchical structure, similar to the structure of the division of labor or even identical to it. The efficiency of production processes means the amount of product that can be produced in a given time [6]. The quantity produced during production can vary significantly—depending on process disturbances, quality level, equipment efficiency and many other factors. The basic factors associated with the production process include: technology, organization of the production process in time and space, appropriate methods and work techniques.

Increasing production and improving the quality of products can be achieved, among others, by increasing its automation, introducing new technologies, and using modern tools. Production efficiency can also be increased by proper reorganization of production processes, both in time and space [7].

Resource scheduling is the process of transforming a project into a work schedule that takes into account available resources. It is commonly understood that, resource planning is limited to the human factor [8].

The adopted work breakdown structure (WBS), within the designed processes, must be subject to ongoing control, assessing the degree of time-cost variances from the assumed plan.

The implementation of each improvement must be preceded with determining the purpose of improvement and a realistic plan for its implementation, by presenting the appropriate tools and methods of action [9], after each change is introduced work standards must be systematically updated, because each change entails an increase in process variability and further implementation of improvements is only possible after stabilizing production.

Supplying production at the construction site with additional working brigades, most often in the subcontracting system, in which an external company offers its own, higher work rate—camouflaged in the piecework rate [10], is justified only in the event that, the contractual penalties for exceeding the contractual period of the investment outweigh the losses resulting from such a reorganization of the delayed schedule.

In turn, the introduction of more registration stages, or even ongoing monitoring of the progress of works at the construction site, allows for faster diagnosis of time variances and earlier use of tools to improve the plan, which would result in, for sure, indicators close to unity—both in terms of schedule and costs—regarding the profitability of works carried out.

The element to pay attention to is the reason for variances, because their source translates into the level of indicators of the EVM method, which is the most known control tool for the implementation of construction projects [1, 11].

The issue of adjusting schedules to existing restrictions, i.e. limits on employment and availability of means of production, is related to the problem of aligning their checking charts [12], which is one of the measures for assessing their quality.

This name is justified because on their basis it can be determined whether the important features of rational organization of works have been retained when drawing up construction work schedules, i.e. whether employment and the demand for basic construction machinery and other means of production will be as evenly shaped as possible. If these schedules show significant unevenness, they should be adjusted accordingly [12].

Depending on the existing restrictions in terms of time and means of production, different models of so called decision tables can be formulated. The scheme of each decision table corresponds to a construction works schedule, realistic to be carried out under specific conditions of the construction site.

Figure 1 presents four variants of the employment schedule. Comparison of the presented versions shows that, the dislocation of production means changes the time reserves of activities [12]. Therefore, it is important to adopt an objective indicator characterizing the degree of alignment of the checking charts.

The dislocation (movement) of means of production between implemented works may relate to the need to comply with the directive deadline for the completion of works covered by the schedule (limitation of time limit), or the inability to increase employment on ongoing processes (limitation of work limit), with the permissible extension of the project completion.

2 Engineering Application

2.1 Description of the Schedule Examined

The schedule of the construction project (erection of the raw steel and reinforced concrete production hall), was used to carry out the production control test with the use of milestones.

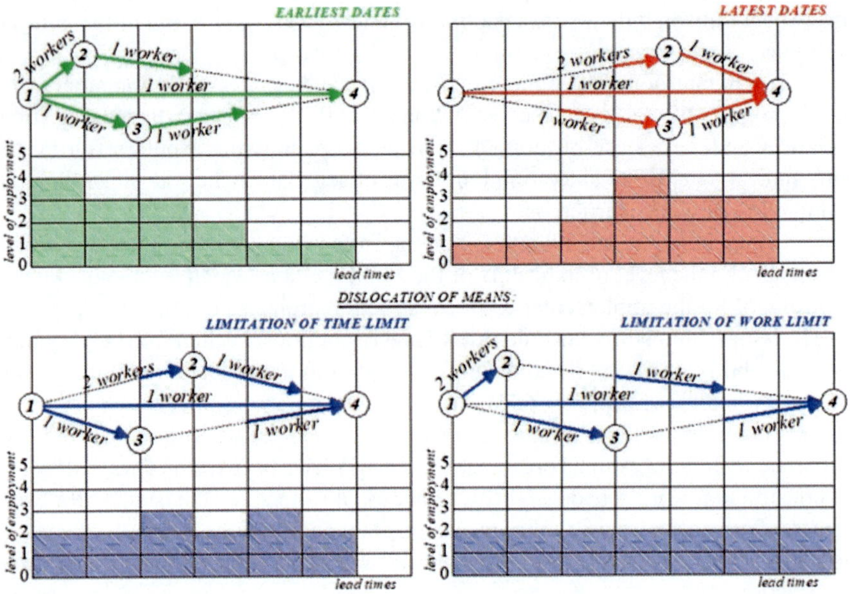

Fig. 1 Network models and employment schedules. Description in the text. *Source* Own study based on calculations of the Planista Max computer program [12]

Scheduling processes were scheduled to be performed without time reserves, obtaining uninterrupted critical path for all assortments of works. The project commencement date was set for October 1, 2015.

The first stage of control over the registration of works at the construction site was carried out on November 16, 2015. After seven weeks of works, a time delay of five working days was found.

By the day of the first imposed inspection of works, the range of processes related to the implementation of the zero level of the hall (activities 1–8) was completed, and the laying of the concrete foundation layer for the reinforced concrete slab was started.

Partial delay in the process in question was two business days for activities 1, 2, 3 (*Site preparation, Humus removal, Excavations for foundations*). Extending the lead time by another three business days was noted for work processes 4, 5, 6, 7 (*Foot underlays, Hall foundations, Foundation insulation*). For activity No. 9 (*Underlay under the slab*), which was production in progress on the control day, an estimated increase in delay was found—up to 7 working days (Fig. 2).

Figure 3 illustrates the analysis of the progress of works using the earned value method on the first day of the report (16.11.2015). The costs incurred for production represent a lower value than costs planned for that day (ACWP < BCWS)—which in a simple comparative analysis may indicate savings.

However, the analysis of the schedule variances (CV, SV < 0.00) and the earned value indicator (BCWP < BCWS, BCWP < ACWP) clearly show the delay of works,

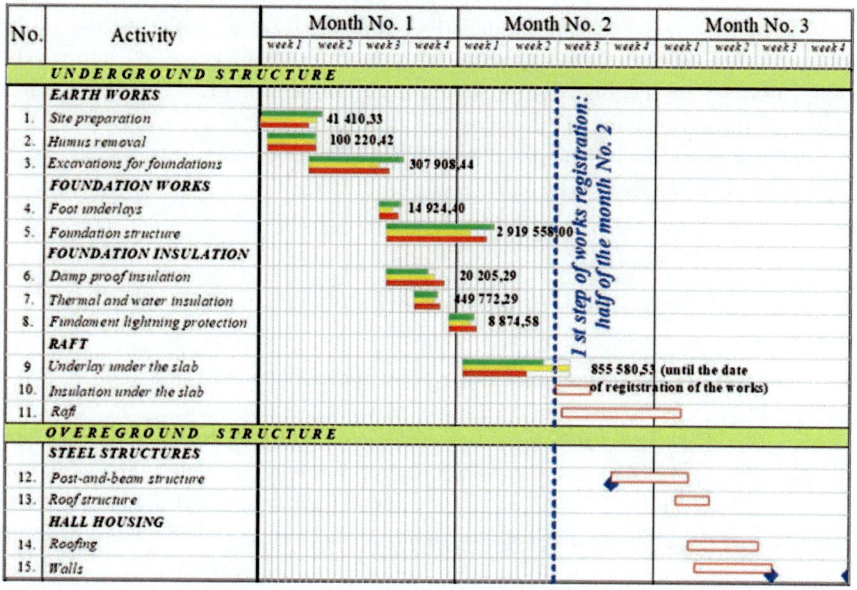

Fig. 2 Schedule of works progress until the first stage of registration: period 01.10.2015–16.11.2015 (activities: green—actual execution, yellow—planned execution, red—costs incurred). Description in the text. *Source* Own study based on calculations of the Planista Max computer program [12]

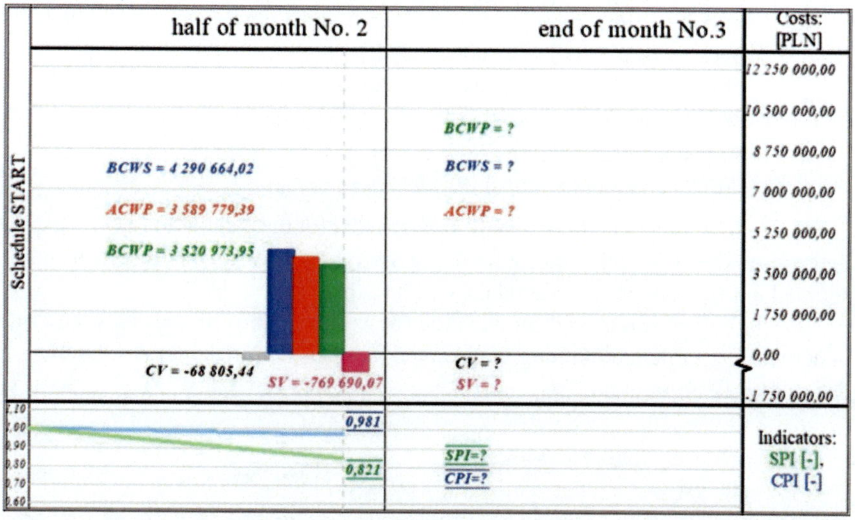

Fig. 3 Analysis of works progress using earned value method: I stage of registration (16.11.2015). Charts: red bar: earned value (BCWP), green bar: incurred costs (ACWP), blue bar: planned costs (BCWS), green line: indicator SPI, blue line: indicator CPI, grey/purple bar: cost/schedule variance; *Source* Own study based on calculations of the Planista Max computer program [12]

while their budget is exceeded. Indicators (CPI, SPI) give a similar picture of the schedule. Profitability of the cost index (CPI: BCWP/ACWP) at 0.981 informs the contractor that, for each capital employed in the amount of PLN 1.00 his return is PLN 0.98 (no surplus, treated as a loss).

Schedule execution indicator at 0.821 (SPI: BCWP/BCWS) illustrates the typical regression of works in relation to the plan.

3 Delayed Schedule Control Model

The total delay of the schedule diagnosed (7 working days) was spread on all activities which were not commenced (10/15), increasing work efficiency on them. The procedure of distributing the supply of activities was used, shortening each of them by one business day—except for the reinforced concrete slab, where the work was accelerated by two days. Supplying activities with additional working brigades was calculated by determining the proposed unit work time expenditure, according to the formula:

$$N(R) = \frac{X}{t \cdot \Pr(1, 2, 3) \cdot G(1, 2, 3)} [\text{m}-\text{h/person/p.u.}] \tag{1}$$

where:

N(R) unit work time input [number of man-hours/1 employee/production unit],
X number of planned man-hours to perform the activities [m-h],
t duration of activities in days, according to the norm [days],
Pr(1,2,3) number of employees servicing the activity on the first, second and third working shift [person],
G(1,2,3) duration of working shift [m-h].

It is worth noting that, the schedule repair model for the calculation of unit expenditure uses the resultant duration of the activity, and not its cost estimate norm, adopted in the catalogs of material expenditure (n_r).

Calculation of the required compositions of working teams for the analyzed series of activities—while maintaining the planned number of shifts and their duration, is determined depending on:

$$\Pr(1', 2', 3') = \frac{X}{t' \cdot G(1, 2, 3) \cdot N(R)} \tag{2}$$

where

Pr(1′,2′,3′) the required number of employees servicing the activity on the first, second and third working shift [person] in order to complete the project on the planned date,
t′ required duration of activity in days [days],

X, N(R), G(1,2,3) as above.

Table 1 presents the results of calculations for the distribution of the total delay, together with an indication of the number of additional production means (workers) required to supply the works in order to complete the project on time.

Figure 4 presents the initial and modified (blue) table of activities of the monitored schedule, with reference to activities No. 10/15—after registration of works on

Table 1 Data on the progress of the works until the first stage of registration: period 01.10.2015–16.11.2.015

Distribution of the total schedule delay	Activities not commenced/as of 16-11-2015					
	No. 10	No. 11	No. 12	No. 13	No. 14	No. 15
Planned duration: t [days]	**5 days**	**20 days**	**11 days**	**5 days**	**10 days**	**11 days**
Normative number of man-hours: X [m-h]	143,225	2,957,404	145,957	82,353	155,780	181,507
Planned number of employees on shifts: Pr (I/II/III) [person]	15/15/0	50/50/45	8/8/0	10/10/0	15/5/0	15/5/0
Planned duration of shits: G (I/II/III) [m-h]	10/10/0	10/10/10	8/8/0	8/8/0	8/8/0	8/8/0
Planned start date	14-11-2015	16-11-2015	27-11-2015	04-12-2015	07-12-2015	12-12-2015
Planned completion date	19-11-2015	08-12-2015	09-12-2015	09-12-2015	17-12-2015	24-12-2015
Distribution of 7 day delay in works: Y [days]	1	2	1	1	1	1
Required duration time: t′ [days]	$5-1=\mathbf{4}$	$19-2=\mathbf{17}$	$11-1=\mathbf{10}$	$5-1=\mathbf{4}$	$10-1=\mathbf{9}$	$11-1=\mathbf{10}$
Unit work time input: N(R) [m-h/person/p.u.]	0.501	0.333	0.500	0.499	0.499	0.500
Number of employees required per shift: Pr (I′,II′,III′)	*<36 >: 18/18/0*	*<165 >: 60/60/45*	*<18 >: 9/9/0*	*<26 >: 13/13/0*	*<22 >: 16/6/0*	*<23 >: 17/6/0*

Source Own study based on calculations of the Planista Max computer program

No.	Activity	Unit pricing (R, S)		Workers at work shifts			Duration of work shifts			Duration of the activity [days]	Activity cost budget [PLN]
		r-g	m-g	Pr1	Pr2	Pr3	G1	G2	G3		
	RAFT										
10.	*Insulation under the slab*	*1 432,25*	*82,67*	15/18	15/18	0	10	10	0	4,77 / 3,98	753 824,59
11.	*Raft*	*29 574,40*	*16 269,62*	50/60	50/60	45	10	10	10	20,40 / 16,92	2 979 338,07
	OVERGROUND STRUCTURE										
	STEEL STRUCTURES										
12.	*Post-and-beam structure*	*1 459,57*	*613,96*	8/9	8/9	0	8	8	0	11,40 / 10,14	207 018,89
13.	*Roof structure*	*823,53*	*273,84*	10/13	10/13	0	8	8	0	5,15 / 3,96	92 732,86
	HALL HOUSING										
14.	*Roofing*	*1 557,80*	*417,20*	15/16	5/6	0	8	8	0	9,74 / 8,85	130 549,61
15.	*Walls*	*1 815,07*	*533,40*	15/17	5/6	0	8	8	0	11,34 / 9,86	143 419,05

Fig. 4 Initial and modified (blue) table of schedule activities after the first stage of registration: Intro-duction of the second working shift to perform the sequence of activities 10–15; *Source* Own study based on calculations of the Planista Max computer program [12]

November 16, 2015 and supplying activities with additional means of production. Increased labor flows were evenly distributed over the duration of individual work shifts in the project.

Figure 5 depicts the distribution of the total schedule delay using milestones in which the required deadlines for completing actions 10–15 have been given feasibility intervals (“*no later than*”).

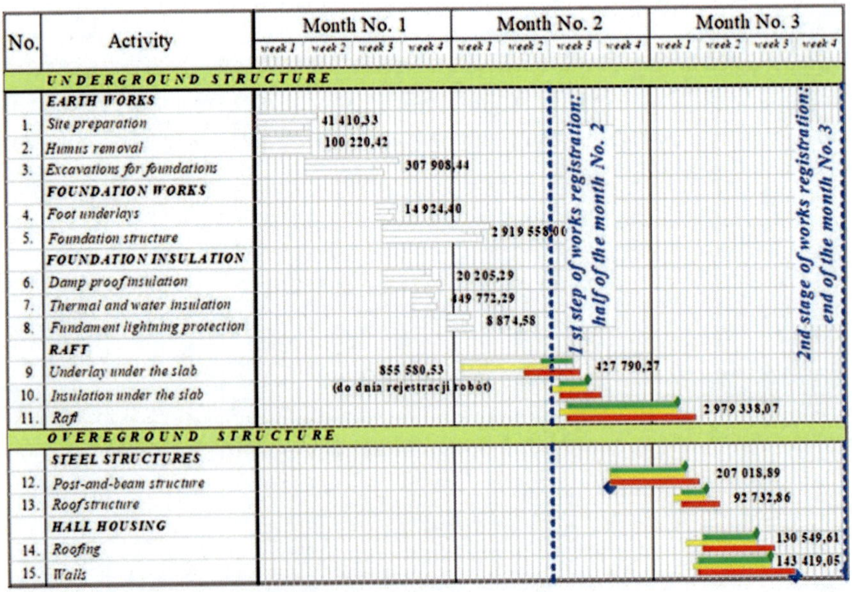

Fig. 5 Distribution of the total delay of the schedule using milestones until the second stage of registration: period 17.11.2015–31.12.2015 (activities: green—actual implementation, yellow—planned performance, red—costs incurred). Description in the text; *Source* Own study based on calculations of the Planista Max computer program [12]

The adjusted durations were recorded with the accuracy of full working days, as per the rules for generating schedules in computer software.

As a result of increasing the efficiency of works for the analyzed activities, the deadline for the implementation of the entire project falls on the planned date of completion of the project, and the costs incurred for their implementation (ACWP) exceed the values of planned costs (BCWS).

4 Analysis of Costs Incurred—Based on the Schedule and Registration of Works

Supplying works with additional working brigades to accelerate the implementation of the project results in increased costs actually incurred (ACWP). Figure 6 illustrates the second progress report using the earned value method (EVM) as at the end of the project (31/12/2015).

The diagnosed delay in the implementation of the project allowed for the early introduction of corrective actions, which "brought" the works carried out on schedule to the "initial path". The increase in the number of employees, however, resulted in an increase in payroll costs, by a total of 11.97%, which resulted in a decrease in the value of the cost performance index (CPI = 0.958), compared to the first stage of work registration. At the same time, the costs incurred for works (ACWP) increased,

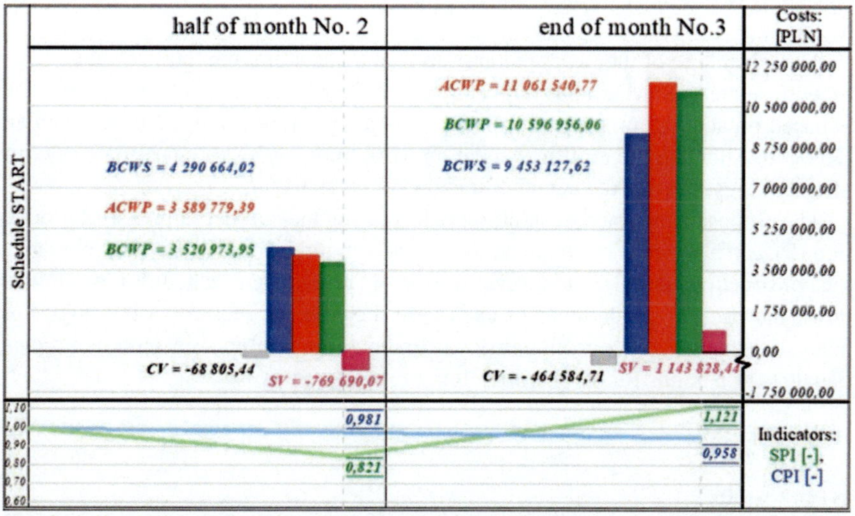

Fig. 6 Analysis of works progress using earned value method: II stage of registration (31.12.2015). Charts: red bar: earned value (BCWP), green bar: incurred costs (ACWP), blue bar: planned costs (BCWS), green line: indicator SPI, blue line: indicator CPI, grey/purple bar: cost/schedule variance; *Source* Own study based on calculations of the Planista Max computer program [12]

exceeding the planned costs (BCWS) and earned value ratio (BCWP) on the day the project was completed. Losses incurred at the construction site illustrate the growing deviation from costs (CV < 0.00).

The acceleration of works above their planned performance is illustrated by an increase in the schedule indicator (SPI = 1121), and the associated positive time deviation (SV > 0.00).

The adopted solution, although increasing the cost of the undertaking, may sometimes be the only option for managers at the managerial level before avoiding contractual penalties for untimely completion of a construction contract.

Therefore, each delay in works execution and forecast of their untimely hand over to an investor requires each time individual economic and financial analysis.

In such an analysis, the comparative element should be the costs of contractual penalties, summarized with the cost of corrective actions accelerating the work above their planned performance.

5 Conclusions from Research and Analyzes

The issue of delayed implementation of construction projects—especially long-term and high-budget ones—is very current.

Their specificity of implementation—among others seasonality, labor shortages, dependence on atmospheric influences, inflationary influences—forces managers to constantly review the assumed schedules and monitor the progress of works covered by them.

The advantage of the proposed approach of distributing the total schedule delay to work sequences, which has not been realized until the work progress is checked, is based on actual, current performance. The adoption of a cost estimate standard (KNR) to calculate the unit time spent on work, in many cases out of date in terms of technology, distorts the actual efficiency of works.

The proposed approach, in addition to increasing the number of means of production (P), can also serve remedial actions, consisting in extending the working day at the construction site (G), and reorganization of "accelerated" schedules, by slowing down the pace of production. In each case, however, it requires a thorough cost analysis consisting of a profitability calculation—in combination with contractual penalties or a bonus for early completion of works.

References

1. Przywara D, Rak A (206) Assessment of time and cost variances of the schedule using the earned value method. Build Mater 6
2. Wypijewski P (2014) Implementation and application of controlling in a large production enterprise. Controlling Manag Acc. Mag. 11:182

3. Dubik MB (2005) Information and decision-making problems in investment activities. Publishing House of the University of Economics in Katowice, Katowice
4. Pająk E (2008) Production and services management. Publishing House P.W.E, Warsaw
5. Brandenburg H (2002) Project management. Publishing house of the University of Economics in Katowice, Katowice
6. Waters D (2001) Operational management: goods and services. Publishing House PWN S.A, Warsaw
7. Nazarko J, Kiełtyka L (2008) IT tools in production management and engineering. Publishing house Difin S.A, Warsaw
8. Kasprowicz T (2002) Engineering of construction projects. W.A.T, Warsaw
9. Kosieradzka A (2012) Enterprise productivity management. C.H. Beck, Warsaw
10. Przywara D, Rak A (2016) Estimating the schedule labor costs against the background of its time delays. Civil Environ Eng 7
11. Przywara D, Rak A Modeling financial liquidity of a construction project in the context of assumed payment deferrals. Opole 2016 Issues of Environmental Engineering in Construction
12. Jaworski KM (2009) Construction implementation design methodology. Publishing house PWN S.A, Warsaw

Time and Cost Variance of Construction Projects Monitored by Bank Investment Supervision

J. Konior and M. Szóstak

Abstract Technical, financial and organisational time and cost monitoring in compliance with a Bank Investment Supervision (BIS) requirements have been presented. Methodology of construction project appraisal for financing and execution professional preparation have been laid out—technical documentation, arrangements, realisation. Engineering, Project and Construction Management (EPCM) approach to investment process and solid, consequent, regular construction projects monitoring executed by professional Banking Supervision Inspector make possible to minimise any faults of inappropriate financing of construction projects. On the basis of the 15-year author's practice of Bank Investment Supervision developed methodology significant construction projects time and cost variance have been identified and measured. The research has been recorded in 536 reports of 40 investment projects in advanced manufacturing, commercial, residential, hotels and apartment housing sectors.

Keywords Time and cost variance · Construction project · Bank investment supervision

1 Introduction

The Bank Investment Supervision (BIS)/Project Monitoring services stand for monitoring of a construction project in terms of risks that may occur [1–3] as well as the financial schedule of all works, in order to evaluate the progress of the works in respect of the loan drawdown provisions.

The scope of the services includes:

- Preparation of Initial Report;
- Preparation of Monthly Reports;

J. Konior (✉) · M. Szóstak
Faculty of Civil Engineering, Department of Building Engineering, Wrocław University of Science and Technology, Wybrzeże Wyspiańskiego 27, 50-370 Wrocław, Poland
e-mail: jaroslaw.konior@pwr.edu.pl

Z. Zembaty et al. (eds.), *Environmental Challenges in Civil Engineering*, Lecture Notes in Civil Engineering 122,
https://doi.org/10.1007/978-3-030-63879-5_16

- Preparation of Final Report.

The main parties involved in carrying out the construction project are:

- Bank/Lender—Financial Institutions considering a grant of an investment loan to finance the objective investment Project;
- Investor/Borrower—Economic entity, acting under the jurisdiction of the laws of Poland which intends to begin the investment Project;
- Engineer/Project Monitor—Company providing services of technical supervision on behalf of Financial Consortium, prepared to undertake monitoring of the progress of works during the investment.

2 Bank Investment Supervision as Source of Research

The objective of executed research was to point out the actual shape and course of the cost and time curve for existing construction projects and to compare it with the planned budget and schedule in the preparation stage of the analysed projects. In order to fulfill the research goals, it was necessary to obtain real data on completed and successfully implemented construction projects.

The main data that is necessary to undergo the research includes: the planned budget and duration of the investment in the form of a work and expenditure schedule, also measurements of the actual progress of the construction process.

This data is developed e.g. in the form of monthly prepared financial statements containing, among others, the percentage advancement and the value of works completed in a given settlement period, the cumulative value of works completed since the beginning of the project, and also the value of works that are still to be carried out [4].

The data for the development of the author's research methodology stems from the 15-year experience of Jaroslaw Konior and the 4 years of work of Mariusz Szóstak, who carry out Bank Investment Supervision on behalf of banks granting investment credits for non-public contracts.

Data concerning 40 construction enterprises was collected, i.e. 536 reports, which were prepared by Bank Investment Supervision inspectors who participated in the course of the analysed projects in a period 2006–2019 [5].

The authors of the article measured and processed a significant part of this data by conducting monthly direct technical and financial inspections at construction sites of carried out projects. The collected data resulted from measuring the current state of the schedule progress and the amount of executed construction works. Cumulative values of the amount of executed works at a construction site constituted the cumulative cost, which determine the course of the S-curve [6]. This course corresponds to the monitored and controlled construction works. The values of the amount and costs of executed construction works have been regularly presented and recorded in projects reports and in BIS reports. The issuing of the BIS initial, periodical and close

Table 1 Summary of the number of analysed construction projects and obtained reports

Category	Sector of BIS projects	Number of projects	Number of reports
A	Apartment houses	14	189
B	Offices	4	58
C	Hotels	9	110
D	Shopping malls	8	113
E	Logistic warehouses	2	12
F	Health centres	1	12
G	Airport developments	1	36
H	Production plants	1	6
A–H	Totals	40	536

out reports has been preceded by internal validation and verification of the coherence of funds that were indicated by the Bank for financing or refinancing [7].

Therefore, gathered data on finished construction projects is reliable, consistent and legible. It can be used to extract typological research samples for enterprises with a similar profile, or to divide them into the categories of construction projects presented below. Assessments of the cost of construction works recorded in BIS reports with a number exceeding 100 may as well be extrapolated to homogeneous populations, which correspond, e.g. to construction sectors in Poland [8].

All the collected data refers to investments implemented and executed in Poland in the period from 2006 to 2020. The summary of the number of reports obtained with regards to the category of the building objects A–G has been presented in the Table 1.

3 BIS and EVM Alliances

Banking Supervision Inspector elaborates documents that enable the mapping and presentation of the actual progress of the construction process. From the monthly reports developed by the BIS, information about the progress of works carried out in individual implementation periods, the values of works carried out in individual implementation periods, and the values of works carried out cumulatively since the beginning of the works, etc. was obtained.

As a result of the performed documentation analysis, a collective data summary was prepared in terms of Earned Value Method (EVM) methodology [9–13]—Fig. 1:

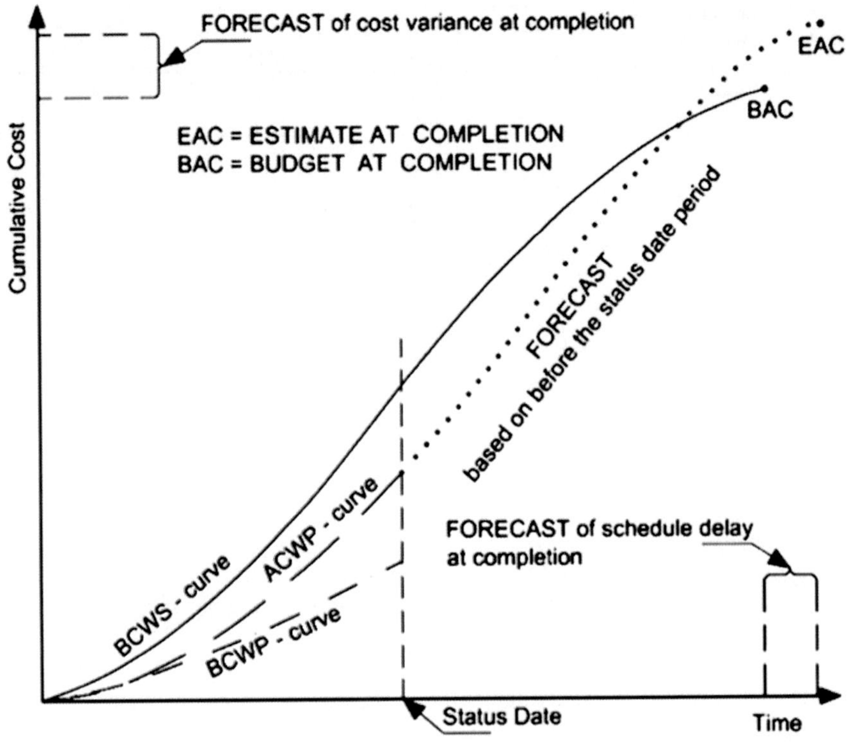

Fig. 1 S-curves representative for earned value method: BCWS, BCWP, ACWP

- the budgeted cost of work scheduled—*BCWS*—determined for each individual examined period on the basis of the investor's work and expenditure preliminary schedule;
- the cumulative value of the budgeted cost of work scheduled—C_{BCWS}—for each single examined period, calculated as the cumulative value obtained by adding the value of the budgeted cost of work scheduled from the analysed period to the value of budgeted cost of work scheduled from the preceding period, determined on the basis of the investor's work and expenditure schedule;
- the actual cost of work performed—*ACWP*—determined for each individual examined period on the basis of BSI reports;
- the cumulative value of the actual cost of work performed—C_{ACWP}—for each examined individual period, calculated as a cumulative value obtained by adding the value of the actual cost of work performed from the analysed period, based on BSI reports, to the value of the actual costs of work performed from the preceding period;
- the actual percentage advancement of work performed—*AC*—calculated as a ratio of the value of the cumulative actual cost of work performed to the total actual cost of the construction project;

- the planned percentage advancement of the work scheduled—*PP*—calculated as the ratio of the cumulative value of the budgeted cost of work scheduled to the total budgeted cost of the construction project;
- the actual percentage advancement of work scheduled—*AP*—calculated as the ratio of the value of the cumulative actual cost of work performed to the total budgeted cost of the construction project;
- the actual schedule variance—*ASV*—calculated as the difference between the actual duration and the planned duration of the project;
- the actual schedule performance indicator—I_{ASV}—calculated as the ratio of the actual duration of the construction project to the planned duration of the project;
- the at-completion variance—*ACV*—calculated as the difference between the total actual cost of the construction project and the total budgeted cost of the construction project;
- the performance indicator of the at-completion variance—I_{ACV}—calculated as the ratio of the total actual cost of the construction project to the total budgeted cost of the construction project.

4 Results of the Research

BCWS, BCWP and ACWP cumulative cost and time curves graphs representative for the most frequently reported group of residential buildings (189 measurements) have been presented in the Fig. 2.

In the third month of construction 20–30% more work was carried out than expected from the fourth month a downward trend of all indicators curves is noticed. Negative deviations from the SV ratio schedule prove that the construction works in that period are usually behind the planned schedule. The trend is easily observed and presented in the Fig. 3.

Construction projects have been carried out in accordance with the planned budgets. ACWP curves coincide with the BCWP curve. In the first months ($0.2 < t < 0.6$, where t is the total duration of the project), the advancement is noticeable more than planned (BCWP > ACWP). This is confirmed by the following graphs in the Fig. 4 presenting the results of the assessment of the degree of construction works progress.

It is noticeable in the second frequently reported group of residential buildings (113 measurements) that no payment was made for the first 3 months of implementation, and ultimately the total cost of this construction project was the same as planned. The deviation cost ratio CV is positive but this is usually due to the payment of the work carried out with a deferred period, not the savings achieved.

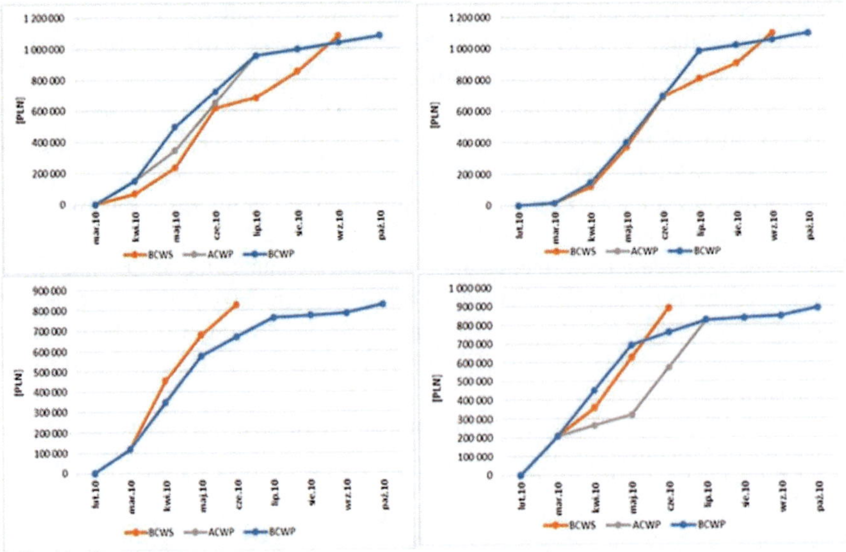

Fig. 2 BCWS, BCWP and ACWP cumulative cost and time curves graphs representative for apartment houses

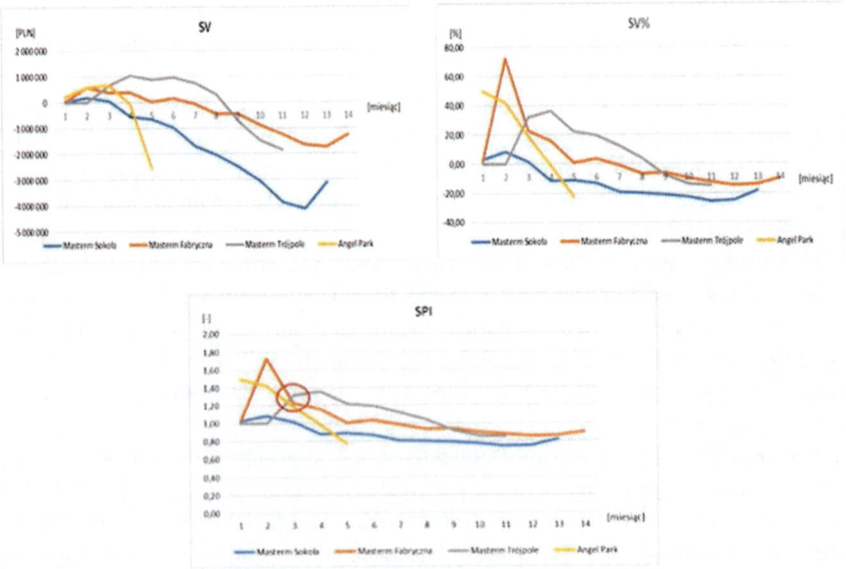

Fig. 3 Variability of time deviations ratios SV, SV% and SPI performances representative for apartment houses

Fig. 4 Variability of cost deviations ratios CV representative for shopping malls

5 Conclusions

The following preliminary conclusions have been drawn as a result of determining time and cost variance of construction projects monitored by Bank Investment Supervision in 40 projects and 536 reported measurements:

- the planned costs resulting from the work and expenditure schedule differ significantly from the actual costs incurred during the execution of the investment; this also applies to the planned time for completing the investment task;
- the average value of the cost performance efficiency indicator proves that the actual cost of implementation is by 2–19% higher than that planned;
- the average value of the schedule performance efficiency indicator proves that the actual implementation time is by 4–20% longer than that planned;
- actual costs, after reaching a half of work and expenditure advancement, are generated at a much faster rate than indicated in the Investor's work and expenditure

schedule until they reach 80%; this is evidenced by the much greater slope of the actual cumulative cost curve towards the time axis.

6 Summing up

The simplest possible models for determining the course of the cumulative construction costs and time curves has been presented in the paper. Practical and easy-to-use model because developed as a result of its own measurements of increasing processing and cost values in cyclical, consistent, verified audit reports of the Banking Investment Supervision.

The Earned Value Method is a simple tool that allows a researcher to monitor the implementation of the schedule and the use of the budget. Several indicators based on scope, timing and cost information can be analysed for the progress of the project, interpret scales, causes and consequences of deviations, as well as estimate the further implementation, including the accurate estimation to complete the project within the budget and the deadline [14].

A reasonable, acceptable level of time and cost variance assessed by EVM approach as a result of BIS analysis for a Bank positive financing cannot be more than 1/5 of the average deviation appraised at each stage EPCM of a BIS project monitoring.

References

1. Konior J (2018) Feasibility study of construction project in compliance with bank investment supervision requirements. J. Civ. Eng. Arch. 12:831–837
2. Konior J (2019) Monitoring of construction projects feasibility by bank investment supervision approach. Civ. Eng. Arch. 7:31–35
3. Konior J (2017) Significance risks evaluation of commercial construction projects. Archiv. Civ. Eng. 65(2):19–33
4. Konior J, Szóstak M (2020) The S-curve as a tool for planning and controlling of construction process—case study. Appl. Sci. 10:2071
5. Multi—Author Work, PM Group Polska sp. z o.o. 2006–2015 and 3EPCM sp. z o.o. 2016–2019, initial, monthly, close-out reporting of bank investment supervision (2020)
6. Cristóbal JRS (2017) The S-curve envelope as a tool for monitoring and control of projects. Proc. Comp. Sci. 121:756–761
7. Konior J, Szóstak M (2020) Methodology of planning the course of the cumulative cost curve in construction projects. Sustainability 12:2347
8. Kapliński O, Konior J (2007) Analytical methods and models in the building project engineering. PAN 249–284
9. Czarnigowska A (2008) Earned value method as a tool for project control. Bud. Archit. 3:15–32
10. Czemplik A (2014) Application of earned value method to progress control of construction projects. Proc. Eng. 91:424–428
11. Kozień E (2017) Application of approximation technique to on-line updating of the actual cost curve in the earned value method. Tech. Trans. 9:181–195
12. Przywara D, Rak A (2017) The time-cost analysis of schedule monitoring using the earned value method. Tech. Trans. 5:57–66

13. Ziółkowska A, Połoński M (2015) Application of the EVM method and its extensions in the implementation of construction objects. Eng. Struct. Technol. 7:189–196
14. Kerzner H (2003) Project management: a systems approach to planning, scheduling, and controlling. John Wiley & Sons Inc, New York, USA